In
Our Hands

In
Our Hands

A Hand Surgeon's Tales of the Body's Most Exquisite Instrument

Arnold Arem, M.D.

TIMES BOOKS

Henry Holt and Company

New York

Times Books
Henry Holt and Company, LLC
Publishers since 1866
115 West 18th Street
New York, New York 10011

Library of Congress Cataloging-in-Publication Data
Arem, Arnold.
In our hands : a hand surgeon's tales of the body's most exquisite
instrument / Arnold Arem.— 1st ed.
p.   cm.
Includes index.
ISBN 0-8050-7179-2 (hb)
1. Hand—Surgery.   2. Hand—Diseases—Treatment.   I. Title.
RD559 .A725 2002
617.5'75—dc21   2002022191

First Edition 2002

Designed by Fritz Metsch

Printed in the United States of America
1   2   3   4   5   6   7   8   9   10

For my wonderful darling, Cindy, who sparkles with creative energy and without whom existence would be drab and uninspiring—thank you for your unswerving support, your love, your confidence in me.

Contents

In
Our Hands

Introduction

In 1833, British anatomist Sir Charles Bell published a book whose premise was that the very existence of the human hand proved the existence of God.

Bell's awe was understandable. Hands possess extraordinary capabilities. They express our fundamental humanity, in our emotions, our music, our love.

As a hand surgeon for twenty years, I have developed a reverence for this exquisite extremity as great as Bell's. Yet, despite their hands' importance, people ignore them and their marvelous abilities. Bell acknowledged this oversight: "We use [our hands] as we draw our breath. . . . Is it not the very perfection of the instrument which makes us insensible to its use?"

Hands command our attention, and their skill is enchanting. Once I stopped for dinner at a restaurant in Bali, where the owner's twelve-year-old daughter danced for the evening's entertainment. As her limber body contorted into impossible positions, her hands became alive telling an ancient story, the *Ramayana,* as richly detailed and full of nuances as a spoken script. Her fingers undulated like ocean swells, bending backward into mesmerizing postures. Pages of narrative became translated into subtleties of movement, hints of meaning. Speech would have broken the spell.

Later, her father told me she began dancing at age five and, by the time she turned seventeen, she'd have to quit because she would be too "old and stiff" to do it properly.

References to hands have crept into our everyday spoken language. We are handy and are proud to survey our handiwork. But if we're heavy-handed and can't maintain our standards, if we have our hands full, we may throw up our hands, wash our hands of the work and hand over the responsibility to someone else. We have to hand it to the inheritor of the job, though. If he or she can handle it, can keep the upper hand, it's not a mere hand-me-down. It's a worthy challenge, hands down.

No single metaphor captures the essence of hands. Hands are central to our humanity—visible, expressive, unprotected. We take it for granted they will always be there because they always have been there. As faithful servants, they are accurate detectors of the world's textures, unerring recorders of the world's hard lessons. As Rodin captured in his sculptures, hands store the memories of our lives, mute yet speaking volumes.

Many simple finger positions, termed "indicative gestures" by Desmond Morris, the cultural anthropologist, communicate symbolic meaning. We cross our fingers in a silent hope for wish fulfillment. We close our fist with the thumb pointing up to signify approval. We never give these casual hand actions a second thought, although their precise meaning is culture-dependent. Making a circle with the thumb and index finger, with the other fingers extended, connotes everything is "A-OK" in the United States. Go to Mexico, though, and the same gesture is an unspeakable obscenity. Making a V with the index and long fingers, Churchill's "V for victory" sign in modern culture with the palm facing *outward,* signals triumph. But the origin of this gesture, with the *back* of the hand facing outward, was a phallic insult. Giving someone "the finger," with the hand turned palm up, extending the long finger straight up with the others flexed, requires no explanation. A clenched fist denotes anger; palms pressed together denotes the spiritual component of prayer; the index finger pointing outward signifies either an accusation or takes the form of an entreaty, as in "Uncle Sam Wants You." These gestures, without words, tell the story of humanity.

Sign language, used by the deaf, bypasses the voice, yet is a rich form of human speech. Hands can read for the blind because members of our species will communicate with each other. Hands can substitute when other senses fail.

Intricate hand anatomy, which captivated Bell, offers a key to the hand's transcendent skill. Presenting hand anatomy in an understandable way is one of this book's goals.

Hands contain a unique network of moving parts unsurpassed in their level of complexity by any other organ. Rich in nerve tissue, the hand is really the organ of information, functioning in everyday use as the brain's extension. Its exquisite sensory capacity, called epicritic sensibility, allows us to reach into a purse or pocket, feel for the dime and pull it out without looking at it. The hand types on a computer, communicating with the best our technology can create in artificial intelligence. Because it moves, it explores objects in three dimensions, reporting the data almost instantly to the brain.

Coordinated, purposeful movement requires a fine interplay between nerves and the myriad muscles whose actions they serve. Draped over a stable architecture of living bone, the muscles become tendons connecting to bone, forming the "puppet strings" that allow the fingers to function.

The great neurobiologist Wilder Penfield mapped the area of the human cerebral cortex devoted to operating the body's parts. The representation of the hands, especially the thumbs, is enormous, larger than any other body part—except the mouth and lips. Penfield's pioneering studies gave credibility to the dominance of hands in evolutionary development, and imparted significance to the idea of "body image."

The special relationship we forge with our hands in infancy is a lifelong interdependence deeper than the shared bond of identical twins. For this reason, loss of hand function affects us in the same way as the loss of a close relative, and we grieve the loss deeply.

Hands are different, distinct from any other part of the anatomy, in another, almost secret, way. We have empowered them to represent us and, in their support, they are competent, unfailingly loyal. In their accuracy and sensitivity, they are almost sentient. But do we know what our hands know? What was the origin of their great skill? What teacher provided them guidance?

When hands become diseased or disabled, normal life stops. When children are born with deformed or missing hands, normal life doesn't even have a chance to begin. Trying to fix the problem is always a challenge. It requires intimate contact with the afflicted person's core personality. My responsibility as a surgeon is awesome, since the first intervention, without restrictive scar tissue to obscure or complicate my efforts, is often the best shot I have to correct the problem. What I accomplish (or don't) determines the individual's lifelong ability to deal with the world. The physical pain involved in the surgery is not as important to the patient as the results. Will it work? Will the fingers be straight? Will they move? Will they feel?

In my general surgical training I've sewn in heart valves, repaired arteries, transplanted organs. But I've found few things in medicine to be as complicated or technically demanding as hand surgery. The anatomy may be (and often is) different from normal. Nerves can branch unexpectedly. There may be tendons that should have disappeared in embryonic life but didn't. Trauma can often disrupt the anatomy with delicate structures torn and fragmented. And because you're always working with an arm tourniquet in place to stop bleeding (so you can see what you're doing), you have a two-hour safe time limit—not much time if there's a lot to do.

During my years of specialty training I went from general to plastic surgery, then on to hand surgery. I wanted to devote myself to more fundamental challenges, issues of psychological distress, problems requiring more technical skill, problems dealing with future self-determination, self-worth, and productivity. But this choice is not without its frustrations. Hand problems strike at the core of personality. People are helpless without their hands. Hands take care of us, earn a living, communicate, write poetry, feed and dress, build roads, make furniture, make love. . . . Hands participate openly in our living history.

It's impossible to re-create the perfection of normal hand

anatomy and function, through I strive to do so. And it's what my patients want, too.

Hand surgeons intervene when people acknowledge they can't do what they need or want to do. For many, like those afflicted with rheumatoid disease or carpal tunnel syndrome, the desired action may seem mundane—holding a pen and writing. But the solution is never simple. It requires navigating a minefield of potential hazards: scarring, inadequate tissue integrity for reconstruction, variations in anatomy, and inadequate blood supply are only a few.

As a hand surgeon, I'm a problem solver. And there are lots of things which can, and do, go wrong. Problems can be categorized into the basic tissues of the hand: nerves; tendons; ligaments; bones, arteries and veins; skin and supporting tissues. All of these can be afflicted by disease, by trauma, or by developmental abnormalities seen at birth. Any of these can be worsened by rough handling of tissues, failure to keep them moist, use of the wrong suture material, use of a destructive cautery to stop bleeding, or other technical details, including exceeding safe tourniquet time. The tourniquet, always the tourniquet. Successful surgical treatment can make an Indiana Jones cliffhanger look like a walk in the park.

Because we don't regenerate body parts, like salamanders, all wound healing depends on formation of scar, "nature's glue," which heals everything but at a potentially heavy price. Tendons, the "puppet strings," must glide smoothly for them to make the hand and fingers move. Scar gets them stuck. Ultimately, all hand surgery involves an endless negotiation toward getting things healed but not stuck.

The stories in *In Our Hands* detail a number of common and uncommon hand maladies and the emotional and psychological fallout that resulted. Each patient I treated had a story. Their stories spoke of grudging acquiescence to reality, of accommodation, of hope. Despite the challenges, these patients found solutions. The lessons they learned come from the complex fabric of human experience.

To me, no hand problem is "minor." I've treated too many patients who have lost or injured their hands to take hands for granted.

I've come to believe that hands are, in a true sense, the Rosetta stone of the soul. One has but to decipher the encrypted lessons locked in their deformities to uncover the wisdom they harbor.

This book is divided into two parts. Part I chronicles stories and anecdotes about patients I've cared for over the years. Part II gives dimension to the hand problems my patients experienced. All the cases described are based on real patients. I've changed names and places to maintain anonymity. Because several patients were treated years ago, I have paraphrased dialogue. However, it is true to the intent, context, and the flow of the narrative.

PART I

Tales of the Hand

I.

A Nice Touch

I never met Muppet-master Jim Henson. But I did meet the germ that killed him. I know it was the same implacable foe, because it left behind its business card: a near-dead human being with both legs and one hand amputated, and the remaining right hand a blackened husk with a few remnants of fingers pinkly peeking through the necrotic char. The encounter had the illusory quality of a dream, spoiled by the rude fact that it was very real. It played like a Hollywood movie scripted by a madman.

There was a time when I relished the schoolboy excitement of being urgently called in to deal with monstrous challenges. In my general surgery days there were the ruptured aortic aneurisms, gunshot wounds, and motorcycle wipeouts to joust with. Lots of blood out and in. Tears and anguish. The sound of wailing sirens and wailing relatives mixed together in a surreal cacophony.

But no longer. Years of sleepless nights in agitated, noisy emergency rooms and tense, hushed operating rooms cured me of all of it. I've traded quantity-of-life for quality-of-life issues. Hand surgery is a gentlemen's specialty. No more standing for hours over a steaming abdomen, draining thick yellow pus or closing bowel. I've come to enjoy more refined problems, predictability in my schedule, and time off with my family.

So I was not happy, not happy at all, when Ted Abelson's orthopedic surgeon called, described the situation in sober tones, and asked me to see Ted at the hospital and assume responsibility for his care.

Responsibility. Now there's a euphemistic term that doesn't do justice to the harsh reality of what I would be facing. A prosperous, athletic forty-two-year-old accountant had, six weeks earlier, developed a flu-like illness, sore throat, fever, and malaise. No big deal, or so it seemed at first. But within thirty-six hours he was near death, in liver and kidney shutdown. Emergency blood cultures were drawn, and he was placed on life support and given massive doses of intravenous antibiotics. But the toxins from the overwhelming bacterial insult had produced a catastrophic complication—his blood was clotting inside his own bloodstream. Infected blood clots, called septic emboli, were being propelled by his still strong heart out into his extremities where they lodged in the major arteries, choking off all nourishment and producing rapid gangrene. Before the infection could be controlled, both legs had to be amputated below the knee and the left arm amputated at mid-forearm.

Responsibility. Who or what was responsible? The microbe? The microbe was streptococcus, a chameleon ready to dole out a strep throat, rheumatic fever, or death—take your pick. The victim? Unlucky genetics, poor host defenses, bad karma. His doctors? It took several days of frantic work just to make the diagnosis of "lethal strep" while keeping him alive. Jim Henson's medical team didn't make it in time to save him. So Ted was lucky—wasn't he?

Responsibility. It's a horrible burden that ages me far beyond my fifty-eight years. Is the burden like the torture of Sisyphus, toiling to push a boulder uphill, nearly reaching the pinnacle, only to have it roll back to the bottom? Is the burden our system of medical care, training the populace to depend on doctors for miracles? Or is it just me, a hopeless perfectionist striving for normal function as I reconstruct hands? Hands that allow us to be human beings. Hands that earn a living, feed and dress, brush teeth, comb hair, build roads, write poetry, make furniture, make love. Hands deformed by arthritis or crippled by trauma. Hands distorted by accidents of birth.

But hands gone? That's a tough one.

As I drove to the hospital to meet Ted I yielded to the temptation of second-guessing what I would encounter. I couldn't help recalling a specific "sick joke" I've carried with me since adolescence.

Sick jokes—tiny parables so obscenely offensive the brain's anger fizzles out in helpless frustration and they become funny. They endure because of a grim truth underneath. This particular one dealt with two boys asking Mrs. Jones if Bobby could come outside and play baseball. "But Billy, you know Bobby is a four-limb amputee." "Oh—then can we use him for second base?" I only hoped Ted himself wasn't succumbing to the same hopeless line of thinking.

So I tried, as I pulled myself up the hospital stairs, to steel myself against the unknown. It would have pleased me to enter the room and find a callous, unpleasant human being, a whiner bitching about the food, the nurses, and the unfairness of life. At least that would save me the energy of identifying with him. It would allow me the extra luxury of remaining cool and indifferent. Yes, Mr. Abelson, life is inherently unfair, it's just a matter of degree.

Once, on the streets of Hong Kong, I saw a Chinese beggar who had obviously sustained third-degree burns to his face and hands. His fingers looked as if they had melted together into a fused slag. His face was a Halloween skeleton covered by thin desiccated leather, the nose like an alpine mountain weathered flat by eons of erosion and wear, the eyes perpetually open and red from dense scarring and complete loss of eyelids. Life is both unfair and cheap in that part of the world and no help was, or would be, coming. I shed the tears his deformed eyes could no longer manufacture and emptied my pockets into his cup.

So what I wanted to meet was a wimp. I wanted to meet someone I fundamentally didn't like, who could ease my burden of guilt, guilt over being whole, by allowing me not to care much for him.

For the first—and only—time, Ted Abelson disappointed me.

His good-natured calm surprised me most. His intelligent eyes still twinkled with what I interpreted as some hidden amusement. I surveyed the wreckage of his body suspended on the hospital bed. His three healed stumps reminded me of a great oak whose limbs had been pruned without mercy, except no branches would ever again sprout from them into new life. We were down to one, or fractions of one, and how to make what was left (or would be when my work was completed) serve all of his needs for the next forty or fifty years.

It was evident from our conversations that Ted had come through an extraordinary experience. What was miraculous was that he was alive. What was marvelous was that he had all his mental faculties, not just unscathed but deepened in an unfathomable way. Like a voyager beamed back in a macabre *Star Trek* episode, all his molecules had been scrambled and unscrambled, and this left him with an inner wisdom sought with a passion by the ancient mystics. Ted had experienced a near death, seen the white light, and, willingly or not, had come back. As I talked with him about the surgery he would need, more pain to go through, more uncertainty, more difficulties, he reassured me. I was talking to him in a parched voice about the wrath of God and he was reassuring me, for Christ's sake! I liked him immensely.

If you're a hand surgeon, toes don't rate but fingers do. Unlike the toes, we don't number fingers, we name them. They are like old friends and each deserves acknowledgment. The opposable thumb is special and in a class by itself. Worth 50 percent of the hand (at least, according to the industrial insurance companies), the rotated thumb is an ironic twist of evolution. Without one, you can smoke cigarettes and pick your nose but not do too much else. With one, you can shoot a pistol or play a flute. I prayed for a thumb.

With gentle care I unwrapped Ted's right hand to take inventory, peeling back the medicated gauze slowly to minimize the pain, and began to perspire. The small finger—black as coal down to the knuckle joint, dry and mummified. So much for proper etiquette drinking tea. The index finger, the same. The ring finger was a little better, but not much, shriveled and lifeless from its midportion to the tip. The long finger, though, had possibilities. Only the tip was nonviable, down to the base of the nail. The key would be to preserve mobility of the joints. And the thumb. Ah, the thumb. Only partial necrosis of the tip on the side away from what would be the contact area with the long and (maybe) ring fingers. It would take three hours of surgery to clean up the mess and complete the amputations but, barring any new infections, new clots, or hidden pitfalls there would, at least, be *something*.

So we took stock of each other, and though he was the one semiclad in typically elegant hospital style, I felt stripped naked under his gaze. Like a seasoned Las Vegas veteran sizing up the table with practiced cool, he processed my explanations without visible emotion, and his questions were relevant and to the point. The stakes were, after all, pretty high.

"Doc," he said, "you have me at a disadvantage."

He was wrong about that, but he didn't know it yet.

"Mr. Abelson—"

"Ted."

"Ted—what do I have? For starters, I want to catch my breath, then I want to take things one step at a time—just as you've been doing."

Ted tried to scratch his nose with the rounded end of his left forearm amputation stump. The same process of ejecting infected blood clots into his circulation wasn't limited to his arms and legs. A deadly swarm of tiny clots had showered into the small blood vessels of his face. Wherever they lodged, the skin perished, giving the impression he had some weird sort of black measles. Tiny islands of dead skin studded his face, and the myriad sores itched as they tried to heal. "Yeah. It's been a while since anything that happens has been up to me. I started plunging headlong down a tunnel more than a month ago and it's been a rough ride, let me tell you. But the light at the end of the tunnel wasn't the proverbial train coming toward me. That locomotive ran over me right at the beginning, chewed me up and spat me out. No—the light in the tunnel appeared when I died in the hospital in Albuquerque."

"Oh, I see." Sure I did. The room was cool, so why was I sweating again?

I longed to find out more about the mysterious light, but at that moment Ted's wife, Janice, and their teenage son came in and the conversation shifted to concrete strategies and immediate plans.

"Doctor," she asked after the introductions, "be frank with us. What are Ted's chances for having useful function with his hand?" A slim, attractive woman with auburn hair and finely chiseled features, she wore an expression of grim resolve as her voice winced with the anticipation of bad news. She had the shell-shocked look of a trauma victim. Dark shadows (rather than makeup) silhouetted her moist eyes and I wondered with a great sadness how she was rewriting her own life story.

"Mrs. Abelson—Janice—whatever remains of Ted's right hand, I'm going to work my damnedest to make it functional. Anything he has there will be useful since it's all he has left. It's the consummate example of the expression, 'In the land of the blind, the one-eyed man is king.'" She grimaced at my choice of words. "My job will be to salvage everything. I'll fight like hell to make it all work. It won't be easy."

"That's no surprise," she said. "Nothing has been so far."

"The outcome will depend on three things: what I can accomplish in the operating room tomorrow, what Ted can do in therapy afterward, and some luck thrown in."

I outlined my analysis for them, my plans for surgery the next day, and hoped I sounded upbeat. We chatted for a while longer. Ted's son, diminutive and frail, hovered quietly in the background. A mute apparition, like the Ghost of Christmas Future—did his demeanor foreshadow this family's destiny? Young boys at his age need their fathers. What, I wondered, would substitute for flying kites or playing baseball? I commiserated with the grief which I suspected had to be feeding like a vulture on all of their future expectations. As Janice massaged Ted's shoulders and thighs, I left.

Most people don't realize it, but the operating room is generally filled with good-natured and, at times, risqué banter. Familiarity breeds casual acceptance. Despite the awesome array of instruments and equipment, to those of us who daily toil there it's not especially threatening, just a place to work.

For Ted's surgery, however, the mood in the room was uncharacteristic. It was somber—partly because he was awake (only his right arm asleep with an anesthetic block) but largely because of the gravity of the problem. Trite conversation or humor seemed not only out of place but rude.

All major hand surgery requires the use of a pneumatic tourniquet, a special type of blood pressure cuff that can be inflated above arterial pressure and safely held there, with the arm squeezed and emptied of its blood, for as long as two hours at a time. You can't repair a Rolex watch at the bottom of an inkwell, and you can't repair hand anatomy with its delicate structures obscured by bleeding. Ted was afraid of the tourniquet. The thought of even the most unlikely complication spooked him, but the infection, like a typhoon blown out to sea, was long past and the risks were negligible. It was my turn to reassure him.

As the operation got under way it was immediately evident that, for the most part, the major demarcation lines separating living from dead tissue seemed pretty clear. That was the good news. The bad news was that scattered through the healthy-looking skin and fat were pockets of dusky, gray, really questionable stuff. The dilemma I faced was that excessive trimming might not only delete parts that were destined to make it but might compromise the coordinated activity of remaining fingers. On the other hand, inadequate trimming that left dead tissue behind would promote pain, retard healing, and increase the risk of infection. I felt trapped. I looked around, but all eyes were on me and there was no white knight standing behind me to bail me out or tell me what to do. So I said a prayer and did the best I could. The general public, convinced by television and the *National Enquirer* that miracles are routine in medicine, would be horror-stricken to learn surgeons are not only human but, with distressing frequency, have to make impossibly difficult judgments based on inadequate data.

The operating room radio had been swiped, and I tried to hum as I worked. At least the air-conditioning was on. I dislike amputating fingers. To discard any part of the hand is a negation of the principles I have studied and worked for. But sometimes you have to. When a lifeboat is dangerously weighted down, something needs to be thrown overboard to ensure the survival of the group. As I began the amputations of the small, index, and ring fingers, I stopped humming.

A remarkable transformation of Ted's hand occurred with the shriveled, black digits removed. Even with two and a half fingers gone, it looked better. A built-in abhorrence of that cadaveric, mummified appearance must be part of our primordial memory, and I was reassured as pink prevailed once again. The thumb and long fingers were for the most part intact, with enough of the ring to make a three-point pinch, with the thumb touching the other fingertips, possible. A far cry from normal, but better than the worst-case scenario. As I applied the bandages I was cautiously optimistic. The collective sigh in the room when the tourniquet was deflated was not lost on Ted, his face shrouded behind the sterile drapes.

"You're pleased?" he asked.

"Yes. No," I said. "I mean, I accomplished my goals and all went fine. What's left should heal OK. But that doesn't mean I have to like it."

"If I've got some fingers to use," he said, "I'll like it well enough."

The next few months were consumed with physical therapy and hope for the future. Ted was fitted with ankle and shoe–bearing below-knee prostheses, and adapted to them expertly. Within a short time he was bouncing around the hospital ward with heartwarming agility, and I swear there was even a bit of a swagger to his step. As I got to know Ted better my already substantial respect for him increased. He began to make serious plans for reconstructing his shattered life, and my visits took on a reflective, if lighthearted, tone.

"Hi, Ted," I said as I came in. "I know what to get you as a Christmas present."

"Oh? What's that?"

"A set of Rollerblades," I said with a straight face. "We can have your prosthetist engineer your BKs to accommodate a whole spectrum of specially modified gadgets. Just snap 'em on. Shoes are boring anyway. When you want to roll right over somebody, a pair of tank treads. Use the blades when you want to make a quick escape."

"Quick escape, huh?" he snorted. "Where were you when I needed you? Albuquerque was where I needed a quick escape. Although I did make an exit there, I must say."

"Ted, you never did get to finish telling me about that tunnel. That was when you went into liver and kidney failure and started throwing blood clots. Please tell me the rest."

For a few moments he was lost in thought. "Okay. As you know, I got pretty sick very quickly. Most of the time I was only semi-conscious, except for the pain. There was lots of pain." He was quiet for a moment. "Then I flipped into a space that was totally serene and calm. I felt like I was moving fast, I could feel the wind on my face. I didn't have any real sense of detail. The landscape around me was indistinct. I sensed the presence of people and got an impression of age, great age, as if I were passing through the world extending back to creation.

"As I barreled along, my surroundings became more luminous. Not more clearly defined, just brighter in an odd way. How can I describe it?" For a moment he paused, searching for words. "Light was coming from up ahead, and no point of origin was visible, but I had an inner certainty that I was approaching a source of immense power. As I got closer, I could feel the energy pulse through me and it made me so ecstatic I wanted to cry with joy. I was pulled toward it hypnotically and I wanted to lose myself in it. I was at peace. I knew everything was okay and I would be fine. I was going to receive all the answers to every question I had ever asked."

Ted's expression was becoming increasingly animated. He looked younger. All of the scabs had come off, leaving a tight layer of healed skin that rejuvenated his facial appearance. A chemical peel, the hard way.

"Then I suddenly got a message: *Go back*. With a jolt, I did an about-face and, before I knew it, I was in the intensive care unit again, in pain, with a sense of loss. The only answer I got was that it wasn't time yet. So here I am. And you know?" For a passing moment, there was a hint of anguish in his voice. "Now that I'm here, I don't know what I'm supposed to do."

"Oh, is that all?" I said. "I don't know what I'm supposed to do, either, but I didn't have to die to find out."

Ted smiled. He had been doing a lot of it in recent days, and it was especially becoming. "Some people are just naturally dense. Try bungee jumping and ask about the meaning of life as you leap into thin air, wise guy!"

"But that's what I'd like to be, Ted, a wise guy. I want to learn from your experience—vicariously, of course. I've read a lot of second- and third-hand accounts but I've never before had a chance to talk to someone who's been down that tunnel and come back. I want to know...."

What did I want to know? Suddenly, I wasn't sure. I had been anticipating this moment, the rare opportunity to learn of the secrets of existence from one who knew, one who'd been there, a traveler who had visited what Hamlet called "that undiscovered country" and returned. But, as Hamlet also surmised, the notion of it puzzles the will. I found myself speechless.

"At least I've learned this much," Ted said. "I know I'm here for a purpose, although I don't know what the purpose is. My presence has a meaning that extends beyond myself. We make plans, but things don't always work out that way." He laughed, his eyes twinkling with the same amusement I'd seen in his hospital room. "One look at me should convince anyone of that. We take the cards we're given and play them as well as we can. But I'll tell you something— I'll never be afraid again."

I met his gaze and saw in his face a far look, a look of pure equanimity, a resolute inner calm bought at an exceedingly great price. A price I knew I was not prepared to pay. Who had whom at a disadvantage now? I felt awed in the presence of a remarkable human being, remarkable in that all the layers of surface veneer had been stripped off, leaving a purity of essence truly inspiring in its integrity. If Janice had nothing else to be soothed by, she could have done worse. Although a few fingers thrown in were nice. A nice touch, no pun intended.

2.

How Sharper than a Serpent's Tooth

Stan Fotrell rides a souped-up Harley-Davidson FXSTC wearing mirrored nonprescription sunglasses, even though he is dismally nearsighted. He continues to smoke the strongest nonfiltered cigarettes he can obtain at any price, despite a vicious cough and a right lung crippled by a gunshot wound sustained in a robbery he bungled twelve years ago. He lives a fringe existence as the black sheep product of a cultured, genteel California upbringing, shunned by his prosperous parents and male siblings alike. He domiciles a meter-long Mojave rattlesnake, which he keeps as a pet in a small mesh enclosure in his bedroom.

Even if I accepted the first three as the idiosyncrasies of an eccentric, the last one makes him, in my judgment, certifiably nuts.

Although I was born and raised in New York City, I've lived for over thirty years in the Southwest. Both harbor hazards to life and limb, and I've learned to respect the local ecology and its denizens and maintain safe distance. I have, several times while hiking with my wife and kids, encountered rattlesnakes on the trail. Never mind all the park ranger talks and nature shows on public television. That awesome staccato warning instills terror when it disrupts the quiet morning air. The knowledge I am not in the snake's normal food chain somehow brings little cheer when I'm a few feet from an aroused rattler and miles from help. The great western diamondback is fearsome, and the timber rattlesnake was not without some justification given its scientific name *Crotalus horridus*.

In high school in Brooklyn, I had a friend who had a small pet boa constrictor which he loved to handle. The reptile was, on most occasions, gentle and placid, tolerant of people and vice versa. But once, at a party, something alarmed it while wrapped around my friend's arm. The strike was so instantaneous, the movement was an almost undetectable blur. One moment the skin was unbroken, the next there were two clean puncture marks on his biceps. My friend scolded the snake, returned it to its cage and rejoined the party with no ill effects, sporting the fresh bite as a conversation piece. The absence of venom makes a big difference.

Stan Fotrell did not bring his snake to parties. While not a recluse or entirely a social outcast, even in the biker community he is considered strange. Not that he doesn't try hard to play the role and fit in. Six feet six inches tall, pudgy and awkward, he sports a mohawk and keeps his black tresses well greased and slick. He drinks his obligatory beer and religiously wears his tattered leather jacket, a pair of vultures with blood-drenched talons emblazoned on the back, when he rides. Macho as he attempts to be, female companionship, even a dysfunctional relationship, eludes him. The snake, however, exhibits less (or more?) discriminating taste and tolerates his presence with aplomb.

Stan's family has, for many years, observed his behavior from afar with shock and dismay. Television and sleazy movies have failed to provide them with comparable role models to study in their less than enthusiastic attempts to understand him. Living a comfortable distance away, they have found it easier to tune him out. His sister Cathy, however, is enough of a rebel to maintain contact with him and act as a liaison. Cathy, a child of the Sixties, occasionally dated guys in college at UCLA who ate raw hamburger and refused to bathe. She could relate to Stan's unkempt appearance, crude demeanor, and almost barbaric lifestyle. She actually likes him.

Despite Cathy's laid-back style, she did have problems with the snake. Though not squeamish or given to emotional outbursts or anxieties, she had great difficulty rationalizing her brother's penchant for unpredictable danger. Stan's habit was to lovingly feed his rattler live rodents, which he captured with some difficulty in the desert, pooh-poohing the risks. He claimed an almost mythical bond with the creature, a friendship bordering on a manic symbiosis. Since it was a reasonable assumption that the infatuation was one-sided, one might justifiably argue that it was predictable he would someday be bitten. The inevitability of it did nothing to lessen his outrage or his feelings of betrayal when it occurred.

A rattlesnake, a typical pit viper, has a poison sac in each cheek, attached by a duct to a sharp-tipped, hollow fang. When the snake strikes, it may or may not eject its venom, a toxic protein that takes some time for the snake to manufacture. If it uses up its venom indiscriminately, for purposes other than killing prey, it goes hungry. One morning in 1975, a Tucson neurosurgeon, a friend and colleague, stepped out of his front door in a bathrobe to fetch the newspaper and was promptly bitten by a sidewinder, a rattlesnake noted for its unusual pattern of locomotion and its mean disposition. While it scared the living hell out of him, it was a "dry" bite, the snake evidently intent on asserting its own territorial integrity with merely a warning nibble and no envenomation. It made its point. No hard feelings, and no serious consequences.

But when Stan's rattler bit him on his dominant left index finger and injected a full load of poison, the cordiality of their relationship was irretrievably severed. We will probably never know precisely what happened. Stan has been deliberately vague, and the snake has remained silent. Was this the product of a simple miscommunication? Did Stan, klutz that he is, slip and accidentally lurch forward with his hand outstretched, inadvertently menacing the creature? Was he, with equal plausibility, fond of abusing or torturing the caged reptile and had it finally had enough, lashing out in self-defense? Or is it conceivable the snake had begun to look upon chubby Stan as a pretty decent meal?

In the final analysis, it didn't matter. The paramedics who arrived at his apartment to transport him pride themselves on their cool composure under pressure. But even they were unnerved by the scene. Stan, like a grizzly, can be imposing when he rears to his full six and a half foot height. There he was, scantily clad, flushed and sweating with a toxic fever, cursing and raging vehemently at his coiled assailant, wanting fervently to attack it but afraid to get close enough to do so.

He offered no resistance, though, to offers of help. Rattlesnake venom hurts like a bitch as it destroys tissue, and Stan desperately wanted narcotic medication. He was taken promptly to the hospital for treatment and his erstwhile "pet" found a new home in the Arizona-Sonora Desert Museum.

For countless centuries, death from rattlesnake bite has been a gruesome affair. The venom is a mixture of potent enzymes that affect every organ system in the body but characteristically zero in on the blood, heart, nervous system, and lungs. Unfortunate victims may vomit or cough up blood, become paralyzed, and ultimately perish in massive circulatory collapse, bleeding from everywhere as their coagulation mechanisms fail. But Stan was spared from such an inglorious fate by fast, appropriate medical intervention.

I first met Stan in the intensive care unit, his sister Cathy at his bedside. He was obnoxious, belligerent, and in great pain. The first two were habitual for him, but the third was not and made him even more ornery than usual. To encounter Stan is not to love him. I hoped I could deal more with his stricken index finger and less with him.

And stricken his index finger was. Almost as if begrudgingly resigned to sparing his life, the lethal venom had contented itself with unleashing its fury on the hapless digit. No match for the poison, the finger died an agonizing death. As I unwrapped the hastily contrived emergency room bandage, a sickeningly familiar odor wafted up to my nostrils. I have seen my share of dismemberment and decay, and their visual memories haunt me. But smells conjure up remembrances also. This one was not the apple pie aroma of a farm kitchen, or damp leaves after an autumn rain. This was the fetid scent of putrefying flesh begging for antiseptic to mask it. The hospital staff had bathed Stan, so his normally overpowering body stench could not hide the message emanating from his left hand. The expression of bewilderment laced with repulsion on Stan's face spoke volumes. I knew he wouldn't like what I had to say to him. Nor would he like me for saying it.

As I worked, he eyed me with a mixture of loathing and fear.

"Now I know who you are," he blurted out.

"You do?"

"Yeah. They told me you'd be comin' around. You're the hatchet man. You're the guy who's goin' to tell me you have to chop my friggin' finger off."

"No, I'm not."

He scowled, his eyes narrowing. "You're not?"

"No," I said quietly, "you're going to tell me I have to."

"Ha!" he exploded. "The hell I am. I've kind of gotten used to havin' fingers. So I think I'll just keep mine, thank you very much. Maybe you should just bug off. Goddamn friggin' snake."

"Have you looked at your finger lately?" I asked.

"Na! It's a little sore. But no more than when I burned it last summer on my bike's exhaust manifold. It'll be fine."

"Why don't we have a look, shall we?" I said as I peeled away the last layers of gauze. It hurt to do that, and I thought he might try to punch me. Cathy, bless her, restrained him.

If something with shock value was what was needed at this moment, Stan's index finger did not disappoint. It looked horrible. Tremendously swollen, dusky gray-black, discolored and covered with blisters weeping hemorrhagic fluid, it looked to be (and was) in an irretrievably advanced state of decomposition. There are keepers and there are goners. In my judgment, this finger was unquestionably a goner. Stan, stupidly, but not surprisingly, disagreed.

"I've seen worse," he lied. "Why don't we just wrap it up an' let it heal. It'll be OK in a few days."

"Stan, stop being a jerk for a change," spluttered Cathy, obviously sickened by the odor of putrefaction and exasperated by her brother's obstinacy. "This guy is right. Just look at that thing. It's stone-cold dead. All the talk in the world won't bring it back or make it work. So why don't you just let him do his job and let's get on with it."

Stan stuck out his jaw, but had no immediate rebuttal. A strange interplay of emotions clouded his expression. One was the distinct aura of a trapped animal. Stan was not a mathematical wizard, but even by his simple computation, it was two against one. Despite his menacing demeanor he was, at heart, a coward who backed down from confrontations. More compelling was his respect for his sister, the only family member or, for that matter, the only living person who, as far as he knew, acknowledged his humanity. Stan knew at a deep inner level it was slow suicide to alienate her or incur her wrath. But his desire to please conflicted with his irrational fears.

"Aw, Sis—I don't want to have my finger cut off. It ain't natural. Besides, how am I goin' to make out at work? The guys'll make fun of me, they'll ride me into the ground. Bein' a lefty is bad enough. They've already given me a bad time about that."

In some primitive societies it is a badge of honor to carry scars. Self-mutilation is practiced by ethnic groups all over the world, though what passes as beauty in one culture is considered hideous deformity by another. Some Native American tribes look upon any variant from normal hand anatomy as repugnant. Members with reconstructed fingers will keep their affected hands out of sight in their pockets and not use them, even if the function and appearance restored after injury were a miraculous triumph of surgical skill. Stan was not raised with such a bias, but he was squeamish enough to be ruled by it. The mere thought of having to live without a part of his hand made him tremble with fright.

Just when it looked as if he would be an impossibly hard sell, Cathy intervened. "Stanley, I want to hear this, even if you don't. So just be quiet and let the man speak." He grunted, but made no objection. I breathed a silent thank you and made my pitch.

"Stan, I agree it's a bummer to have to lose your index finger. But you've already lost it. Our noses know it. Leaving it to fester even more isn't going to solve anything. Dead meat is a setup for infection that could jeopardize your whole hand. You don't need or want that." His eyes were riveted to the floor, but his fidgeting told me I had his attention.

"What does the index finger do that's unique? You can pick your nose with any finger, including your thumb if you have to, so that doesn't count. You can stir coffee with it, but a spoon works better. So what's so special about the index?"

Stan looked forlorn. It was becoming increasingly clear that he equated loss of a body part with loss of manhood. Behind that bulldog facade was stark, naked terror rooted in ignorance and inadequate coping skills. I desperately wanted to find some common ground for communication with him but didn't know where to begin. As luck would have it, Cathy provided the link.

"Doctor," she intervened, "my brother is a good person deep down. He's more sensitive than he seems, although he does some things that are, well, you know, kind of weird. He needs some reassurance that things are going to come out OK. Frankly, so do I." The foul aroma of a dead finger was, quite understandably, not reassuring. She gazed at Stan with the heartwarming compassion of a Mother Teresa.

Stan's expression remained a mask of grim bravado. I was about to speak when I noticed, at the corners of his eyes, small tears.

In that instant, I was overcome with remorse. I had taken his facade at face value and responded to his stubborn boorishness with impatience, thinking him an adult. But here before me was a hurt, frightened child in a grown-up body. He was doing his best to sustain the illusion of a comic book hero and failing, as he had failed in so much else in his life. Stern indifference was the last thing he needed.

"Stan," I said gently, "the index finger does two things that set it apart from the other fingers. First of all, it points. Look at your other hand and I'll show you."

Grudgingly and a bit clumsily, to avoid pulling on the intravenous tubing, he raised his right hand from the bedsheets.

"First make a fist. . . . Good. Now straighten your fingers." A scantily clad maiden tattooed on the back of his wrist shimmied invitingly as he complied. "There's a big muscle on the back of your forearm attached to tendons that straighten or extend all of the fingers together."

"Now make a fist again and straighten your index finger.... Great! You can see that it has the ability to extend or straighten by itself. It can point when the other fingers are tightly flexed. That's because there's a tendon, separate and independent from the others, which does that. The little finger has one, too. Try it."

With great delicacy, he lifted his pinky away from the others, like a prim dowager at a formal tea. The action seemed ludicrous in this setting.

"Wonderful! The independent extensor tendons give these two border fingers special dexterity the other, central fingers don't have. To prove it, make a fist and try to straighten the ring finger by itself."

He tried, and couldn't. "Fine," I said. "That's one unique feature of the index finger. Now press your thumb against the side of your index finger and see what happens." He followed my instructions like an automaton, not yet getting into the spirit of this crucial lesson.

"See that big muscle attached to the base of your index finger tighten up? That muscle, called the first dorsal interosseous muscle, makes your pinch power strong." I scanned his face hoping for a glimmer of enthusiasm. Nothing yet.

"All of this gobbledygook anatomy is important to you, Stan. It means a lot for the future function of your left hand. You're left-handed, and I know how anxious and upset you are about how you're going to manage when this is all finished." He shifted his huge bulk in bed to see me better, and his lips parted speechlessly. I went on.

"If all goes well, with no great surprises in the operating room, I plan to take that independent tendon and that strong muscle, save them when I amputate your index finger, and reattach them to your long finger. What that will do is make your long finger work like an index finger. The long finger will be able to straighten independently and generate greater pinch power. What's more, the reattachment will 'indicize' your long finger and actually make it feel like an index finger. That's because there are nerves in the tendons and muscles that do that. Your hand will be narrower than before, and a little less strong, but you'll manage fine. Really, you will!"

Stan closed his mouth and stared at Cathy with a glance that reminded me of Warren Beatty staring at Faye Dunaway in the final scene of *Bonnie and Clyde*, just before the machine guns opened fire on them. It was a look of grim resolution, a fatalistic acceptance of whatever the future held in store. But the plan, to garnish true acceptance, required Cathy's seal of approval. She gave it.

"Doctor, I think I am speaking for Stanley when I ask you to go ahead with the work that has to be done. He's been through some tough situations before—he'll weather the storm." Stan's left hand unconsciously strayed to the long scar on his chest.

"Stanley—Cathy—time is working against us," I said. "It's hard sometimes to get into the operating room quickly, but I think, under these circumstances, the staff will be accommodating." Diplomat that I am, I was loath to add my best guess that the ward nurses caring for Stan and his malodorous finger would be willing to pay under the table to get him off their unit STAT.

"Stanley? What's it to be?"

"It won't hurt, will it, Doc? I don't want no pain." He coughed, and grimaced slightly.

"No, Stan," I said deliberately and with great solemnity, "you'll feel no pain during the operation, and surprisingly little afterward. It's nothing like getting your chest cracked."

His sad eyes enveloped me as he nodded silently, and I noted with mute satisfaction that we had finally reached a level of understanding.

In every operating theater, as in every organization populated by people, there is a functioning grapevine. Word spreads quickly. It spreads especially fast in a hospital setting. The mysteries of disease hold a special fascination for most of us, unusual ailments particularly so. Ghastly, messy problems seem to exude a special panache that rivets the attention more than a murder mystery. So it was no surprise that the operating room staff—bosom buddies, after all, with the ward staff—knew a great deal about Mr. Fotrell and anticipated his appearance with morbid interest.

My regular scrub nurse worked the night shift for many years and has seen more than her share of trauma. She is, therefore, some-what sanguine about blood and guts issues, and it takes a fairly grotesque injury superimposed on a memorable personality to impress her or tweak her curiosity. Stan met the criteria. Acknowl-edging a real challenge, she grunted appreciatively at his physical size, tattoos, and hostile demeanor before his bandages were removed.

"Hi, Mr. Fotrell, my name is Karen, Dr. Arem's scrub nurse, and I'll be helping him with your operation today. Why don't you just slide over onto this bed?"

Stanley glowered at her from the sparsely cushioned gurney that had wheeled him into the operating room. "My arm's numb, but the rest of me's like a live grenade. I ain't movin' a twitch till I get guarantees this won't hurt. I don't want to feel no pain. I can't face it."

"Charming," Karen whispered to me quietly, her surgical mask hiding the movement of her lips. "How much will it cost me to get you to forget the arm block and just put him to sleep?"

"Karen," I said loudly, with clear emphasis directed toward our belligerent patient, "Mr. Fotrell is just expressing what any of us would feel in similar circumstances. We're all scared at times. We're all afraid of pain. We need reassurance. It's only natural. Isn't that right, Stan?"

Once again, Stan was trapped. Sheer machismo made him reluctant to admit the simple truth: he was terrified. But having got-ten this far, surrounded by the sterile trappings and cold steel of a sur-gical theater, he badly needed a kind word.

"Well, I don't need no mollycoddling or hand-holding. I ain't thrilled about bein' here, but I got talked into it. So as long as we're doin' this, I just want to be sure that you're up to doin' your jobs, is all."

"That anesthetic block the anesthesiologist used on your arm will last for three hours or so," I commented dryly. "During the oper-ation, you'll feel nothing at all. When I finish the surgery, I'll inject all the deep tissues and skin edges with long-acting local anesthetic which lasts for seven hours. Back on the ward, you'll start taking pain medicine as the anesthetics wear off. As long as you keep your hand elevated above your heart to minimize any swelling, it won't hurt much. You'll be comfortable enough. But you'll be awake for the duration of the surgery—your bum lungs wouldn't appreciate a general anesthetic—and you'll have to put up with our bad jokes. You'll also have to listen to classical music, not country and western, on our radio. I'm doing delicate surgery, not branding cattle."

Stan's sour expression reflected a combination of resignation and relief. I somehow had the distinct impression he would give us no further trouble. For her part, Karen seemed disappointed—I think she was a little bored and scrapping for a good challenge.

The surgery went precisely as I had outlined. Fortunately, the demarcation line between envenomated and healthy tissue was well defined and the tendon and muscle I needed to transfer were out of harm's way, minimizing risks of infection or poor healing. For his part, Stan said nothing at all throughout the procedure, a pleasant surprise to all of us. But as I applied the final bandages he became animated once again. With the operation completed without a trace of discomfort, what little courage he had, having deserted him, returned to pump him up with a mockery of virility and bluster.

"I suppose you had a gay old time telling your friends about my snake."

"I don't need that kind of publicity, Stan," I responded. "Actually, the less said, the better. I don't particularly relish the thought of having to defend you before an inquiry of the Society for Cruelty to Animals."

"Cruel! You think I'm cruel? I'm the one who gets bit, I'm the one who gets his friggin' finger cut off, an' you say I'm cruel. Great humanitarian you are. Hah!"

Stan had shifted his semi-outraged hulk onto the gurney and, before he could say more, the transportation crew whisked him back to the surgery holding area. I elected to face him again in his hospital room. Cathy was there, looking relieved to see him his usual feisty self with a clean bandage. She did not comment on the visible presence of a thumb and only three fingers.

"Doctor," she said, "Stanley was so impressed with the effectiveness of the arm block—he says he never felt a thing!" Stan clenched his teeth tightly together and gazed up at the ceiling. "He's always been concerned about physical pain. Some people seem able to tolerate it pretty well, but not Stanley. I guess he has a low threshold or something."

"Nobody likes or looks forward to pain. In that respect, Stan's no different from the rest of us. But in spite of our best efforts, we still get hurt. We trash our bodies and have to be put back together. Our scars prove it." Stan absentmindedly stroked his chest scar. "In fact," I ventured, "one distinctly unpleasant experience can set the stage for a lifetime of fear, fear of repetition of the experience. It's too bad, really, because that fear is often groundless. But not always. It usually hurts to get your chest opened—doesn't it, Stanley?"

Stan's head rotated slowly until his stony gaze enveloped me. "Yeah—it hurts. It hurts like hell. You should try it sometime. It'll give you somethin' memorable to mouth off about."

"I'll do that—next time I get shot robbing a liquor store. It was a liquor store, wasn't it?"

"Naah, where'd you get that? It was a parts store for my bike, is all. I missed my paycheck 'cause I was out sick, an' I needed a replacement gear chain for my Harley. I would've come back to pay the guy, I swear. He didn't have to blast me. Jeez, that was no fun! Some people have no sense of perspective. They got the bullet out in surgery, though—along with a piece of my lung."

Stan was staring past me at the open door to his room, and his expression abruptly changed to one of frank astonishment and bewilderment. For a moment, as his mouth fell open with a sharp intake of breath, I thought he was having a flashback to that traumatic event and reliving it. Cathy broke the spell. "You have visitors, Stanley. Hi, Mel. Hi, fellas."

"Hey, big guy, what's the good word? I hear you been gettin' chewed up and then carved up. Sounds different, anyway." The grinning apparition who had suddenly materialized in the doorway was a striking figure, tall and lanky with a white handlebar moustache worthy of Wild Bill Hickok and wavy white hair tied back in a long ponytail. His filigreed boots, worn jeans, cowboy hat, and tan leather vest completed the picture of a man who took things in stride—long strides, at that. Behind him, three other men, all in work clothes, crowded the corridor. Stan's face remained a frozen mask of amazement. Since he seemed incapable of introducing his friends, Cathy did the honors, beginning with the cowboy.

"Doctor, this is Mel Parrish. He's Stanley's boss at the shop." The others were Frank, Tyrone, and Beaner. A motley crew, but sincere in their concern for Stan.

"A pleasure, Doc. Me an' the other guys've been worried about Stanley here. First, he don't show up at work. Kind of unusual for him. Then we get a phone call from some hospital person wanting to verify that Stan had a job. We figured we better check it out. An' lookie here! We got a regular invalid on our hands."

Stan's lips moved wordlessly for a brief time. Then he stammered and became audible. "I can't believe you all came here to see me. I can't believe it. Who's watching the shop, if you're all here?"

Beaner, a balding, muscular man with enormous arms and a sunny smile, fielded that one. "Tyrone's wife, Phyllis—she had some free time from the laundromat. Of course, if anyone comes in with an emergency cooler duct repair, she has a problem."

"What's the matter, Stan," prodded Mel, "can't you believe that anyone cares enough to come an' see you in your hour of need? I admit you ain't lovely to look at, but you always do your share of the work, you do it OK, an' you mind your own business. What more could anybody want? Besides, we're a team. We gotta stick together. You're a person havin' a tough time of it right now. It'll all come out all right, you'll see, but you need some support. You'd do the same for any of us, right? You'll be back at work before you know it."

In truth, Mel had analyzed the situation flawlessly. Stan's mouth quivered and small tears appeared again at the corners of his eyes. "But how am I gonna do it, Mel? My index finger is gone for life. I'm a gimp, a cripple. How am I gonna work?" His body sagged into a posture of helplessness.

Mel fixed him with a sharp stare. "You haven't been too observant, friend. Have a look at Tyrone and Frank. Boys! Hold up your hands for the gentleman."

Obligingly, the two sheet metal workers did so. Frank's left ring finger had been amputated just beyond the knuckle joint many years before after a punch press injury. Tyrone had lost the tips of his right thumb and long fingers on some razor-sharp galvanized steel as an apprentice in Mel's shop, a year before Stan had started working there. Both men grinned as they displayed their craggy, callused fingers and palms, secure in and reliant on the serviceability and durability of what remained.

"You never noticed that, did you, Stanley? Right under your nose every blessed working day, an' you never noticed. But why should you? These guys don't moan an' bellyache about it; their hands work fine, they can do their jobs, an' that's just the way it is. You'll work, too. And you'll put all this crap behind you, an' go on with livin'. That's just the way it is."

Stanley had pulled himself up to a nearly military attentiveness. No longer interested in his chest scar, he registered emotions I hadn't previously seen. Was it hope I saw in his face? Self-respect? Affection for those who had shown him their own brand of love? Cathy, who knew Stan so well, could clearly see the change, and her face emitted a sober radiance.

"Stanley," she commented softly, "I've been trying for years to convince you that you were an OK person, worthy of caring and concern by others. But you wouldn't buy it. You kept doubting yourself, putting yourself down. Maybe now you'll accept that what I've been trying to tell you is true. These guys being here is proof of it. Sure, you've had some hard knocks, but like Mel says, that's just the way it is. And you've made something of yourself—you've become liked by your friends. In my book, that's as worthy an accomplishment as you can make in this life."

Stan was quieter and more pensive than I'd ever seen him. He looked at Cathy and each of his colleagues searchingly, took a deep breath, and coughed. Then he looked down at his left hand, nodded absently and nibbled on his lower lip. After a minute, he exclaimed, "Goddamn friggin' snake; those nature boys at the museum better be feedin' it decent meals, like I did. At least, the snake was true to its nature. It didn't fink out, it didn't back down an' it gave as good as it took. It was respectable. I don't like sayin' it, but I think that's worth imitatin'." And no one in the room contradicted him.

After five minutes of small talk, Stan's friends murmured apologies and returned to work. As soon as they had left, Stan bubbled conversationally. He seemed brisk and animated. "Doc, how much longer do I gotta stay in this place? I got things to do."

"You can check out now, if you want to, Stanley," I reassured him. "You'll have to come and see me in nine or ten days to have the stitches taken out, and you'll have to be in a cast for a few weeks to protect the tendon and muscle reconstructions I did. After that, we'll see about some therapy to strengthen your hand before you can realistically go back to work."

Stan brooded a bit and looked carefully at me and Cathy. "You mean that, about me workin'? If that don't beat all."

"Take him home, Cathy. I have a suspicion you and the rest of your family have a lot of catching up to do where Stanley is concerned. If you all want to."

Cathy looked tired but alert, as if she sensed that her unique role in this drama was moving along well, but was far from over. "Doctor, can I call you if there are questions?"

"Cathy," I said, "it would amaze the hell out of me if you had all the answers, with or without my help. Sure you can call me. But there are strings attached."

A hint of amusement crossed her smooth but careworn countenance. "What did you have in mind?" A California girl, born and bred. She had, I surmised, heard that line before.

"I want progress reports," I said earnestly and with great conviction. "I'll want to know what Stan is doing, how his hand is working, and how he is facing the future. The rest is none of my business, if you want it that way. But I feel justified in asking for that much. Stanley?"

Stan smiled a cautious smile. "I ain't makin' big promises. But you ain't askin' for a lot, either. It'll be as you say." He squirmed uneasily, seemingly happy to be getting off without major commitments. Yet even a simple thank you was more than he could muster. I wasn't greatly surprised. For all his massive grown bulk, Stanley was still a work in progress, with edges rough enough to abrade cement. Someone else would ultimately have to smooth them out. Cathy? For her sake, I hoped not. But I was sorry he didn't get to tangle with my scrub nurse, Karen. She would have given him a run for his money.

3.

A Chance to Cure

Marva Stowburn was as mild-mannered as Clark Kent. A white-haired, seventy-eight-year-old granny with a soft, pleasant smile bestowed on everyone, she was demure and inoffensive. Her medical complaint seemed straightforward enough, neat and clean and anatomical. It was perfectly natural to take it at face value and try to evaluate and solve it in the standard way.

After a lifetime of heavy use, Mrs. Stowburn's large hands were arthritic and painful. The public, bombarded with television commercials extolling the analgesic virtues of pain remedies and "arthritic formulas," has come to believe all hand pain is arthritic until proven otherwise. This is not true. There are more non-arthritic causes of hand pain than arthritic ones. Osteoarthritis produced by wear and tear is a specific diagnosis requiring some key features for verification: appropriate complaints, visible changes in the affected joints on physical examination, and abnormal X rays confirming loss of cartilage with narrowing of the joint.

Using these essential criteria, Mrs. Stowburn indeed had arthritis—bad arthritis, involving the thumb, index, and long fingers of both hands, a little worse on the dominant right side. In fact, she was increasingly limited by it. No longer could she sew and crochet into the late night hours. No longer could she knead dough and make the famous dark bread which had, for years, been her special trademark. She knew she was getting older and accepted most limitations with equanimity.

But when she found herself completely unable to flex the tips of her right thumb and index finger, she became alarmed. A confirmed "righty" since childhood, she was distressingly reliant on function of her dominant hand, and this latest infirmity was simply more than she could tolerate. Where will it all end? she wondered. Unable to drive because of a chronic back problem, she tapped a neighbor to ferry her to my office. Years ago, I had treated this neighbor's carpal tunnel syndrome, with a happy outcome. So I had the advantage of vicarious trust to ease the introductions and engender some confidence.

She smiled pleasantly at me, and her brows knitted into an expression of forlorn worry. As an intern I learned this was known as the "omega sign," since the eyebrow wrinkling took on the configuration of the Greek letter *omega*. She pursed her lips.

"This problem is driving me stir-crazy. About four or five months ago my fingers stiffened up, even more than usual. I've been prone to a little rheumatism, but I could always get by with some aspirin and a sinkful of hot water. Lately, though, it's just not working. My fingers won't bend and they hurt. And my skin is bruising to beat the band. Just look at my poor arms, all black and blue."

She held out her arms for me to inspect. The technical term for black-and-blue marks is *ecchymosis*, caused by bleeding under the skin, and she had some beauties on her forearms and hands, probably due to age-induced fragility of the blood vessels combined with the bleeding tendency caused by aspirin.

"Does the aspirin help enough to be worth the bruising?" I asked.

"I think so, but I'm not sure. I'm at my wit's end. I don't know what to do anymore." She sighed. "I once had pretty hands. Now look at them—knobby and awful."

The DIP joints, the ones just below the fingernails of the index and long fingers, had unique bumps, called Heberden's nodes, absolutely characteristic of the underlying arthritic problem. Even without X rays, the diagnosis of wear-and-tear osteoarthritis was clear-cut.

But there was more than just this. Her right hand was oddly

rigid, as if frozen into a set position as assuredly as if a sculptor had carved it this way out of stone. Both the thumb and index finger were held straight, at almost a ninety-degree angle to each other. Most peculiar. She had arthritis, yes. But that didn't explain the frozen position. In fact, her joints were passively mobile. I could take them gently with my hands and bend them. Only she couldn't.

I mentally began to take stock of the possibilities. If you like human anatomy, hand surgery is a delightfully anatomical specialty. When things go wrong, making it all right again often rests in knowing how this wondrous appendage is put together. The constancy of anatomy also provides a certain reassuring consistency in an uncertain world. It's something you can count on.

Or is it? Unfortunately, there are big holes in this smug line of reasoning. Variations or aberrations of normal anatomy are distressingly common. Things aren't always what they seem. You have to know not only the normal patterns but the abnormal ones as well to make sense out of many problems.

There are nerves out of position or going where they don't belong. There are muscles that are duplicated, in the wrong place, or powered by the wrong nerve. There are structures that should have disappeared at birth but somehow persist into adulthood, creating a fine mess. When one adds to this confusion the unpredictable effects of trauma, illness, and individual variability, it's a wonder anybody can diagnose anything.

But you have to try anyway, and you have to start somewhere. The median nerve is one of the most important nerves in the upper extremity. Its sensory branches carry touch information from the fingertips to the brain. Physical compression of these branches at the wrist creates the familiar—and infamous—carpal tunnel syndrome. The nerve's motor divisions in the forearm run to a diverse group of muscles. One of the branches is called the anterior interosseous (literally "between the bones"). This nerve powers the muscles that flex the tips of the thumb, index, and sometimes the long, fingers.

Could Mrs. Stowburn have an anterior interosseous nerve syndrome, a localized and very specific compression of this nerve, thereby producing paralysis of the affected muscles and resulting in her odd hand posture?

For starters, this certainly seemed reasonable. The condition was first described in detail in 1918, and hundreds of articles about it have been published in scientific journals since then. Never mind that there are at least eight different anatomical abnormalities that can cause it. Working the details out would be a surgical exercise for later, if we ever got far enough. The first order of business was to confirm the diagnosis.

Happily, there are noninvasive ways to do this, with no cutting required. Armed with the right equipment, a neurologist or physiatrist experienced in the technique can stimulate the nerve with a low-voltage current, insert fine needles into the affected muscles, and, as the nerve is stimulated, take recordings from the muscles (the EMG, or electromyogram). If there is a compression of the nerve, recordings of muscle activity seen on an oscilloscope hooked up to the muscles may be abnormal. If the nerve is compressed, the rate with which the nerve conducts an electrical impulse, or "nerve conduction velocity," slows down at the point of compression, and this also can be accurately measured on an oscilloscope. The information derived from such testing creates a fingerprint of the lesion.

I went over all of this in detail with Mrs. Stowburn. She wasn't thrilled about the needles (who is?), but she was distressed enough about her functional predicament that it seemed a small price to pay, so she agreed. She was cautiously enthusiastic when she left my office.

She was despondent when she returned. As I reviewed the EMG/nerve conduction study and the neurologist's interpretation, I could see why. The results were equivocal, borderline. Maybe she had nerve compression, or maybe she didn't. Maybe the stiffness in her fingers was due to her arthritis, or maybe it wasn't. And the bottom line was an even bigger uncertainty. Would surgical exploration benefit her? Did she really have compression of her anterior interosseous nerve, with paralysis of the flexors to her thumb and index finger based on this, only we couldn't prove it unless we surgically released the nerve and watched everything return to normal? Was the study inconclusive because of some anatomical oddity of her arm? Or were we simply facing the limits of accuracy of a mechanical test which, after all, is just a big artifact and the subtleties of human biology had outsmarted us once again?

No question, there are "cutting surgeons" who would not have hesitated recommending surgery. After all, as we were indoctrinated in our training, "A chance to cut is a chance to cure." Or, "If you want to save a life you have to use a knife." Or, "If you want to heal, you need cold steel." These absurd little homilies have woven themselves into the fabric of thinking in modern surgery. One of the best ways to put them in proper perspective is to ask, "What would I want if it were me?" Would I want a surgical operation? As a surgeon, putting yourself in the patient's shoes and adopting a "do unto others" philosophy keeps you honest. Poorer, but honest.

So when she and I sat down to discuss the results, we both wrung our hands in frustration—I literally, she figuratively, since it would have been painful for her. My office is a private home that I bought, remodeled, and filled with my oil paintings and stained-glass projects. The waiting room (the former living room) is graced by a flagstone fireplace and the ambience is more relaxing than a coldly sterile, typically antiseptic office. My exam room is a wood-paneled bedroom whose sliding glass door opens onto a shady, tree-lined patio brought to life by frolicking birds and squirrels.

We sat at my exam desk facing each other, and I studied her face. The omega sign was especially prominent, as were the careworn lines of a lifetime of—what? Abuse? Hard work? Financial troubles? As I looked closely at her, I realized I didn't know this woman at all. And I began to get a strange feeling. An odd sense, skirting the fringes of my awareness, that I was missing the boat somehow. Something vague and indefinable, a half-formed notion that I ought to proceed differently.

Sitting with Mrs. Stowburn in my exam room, with the birds chirping and a serene sunlight filtering through the window, I was strangely motivated, intuitively compelled, to ask her questions about her childhood. Why, you might ask, should the childhood of a seventy-eight-year-old woman have any bearing on her newly developed nerve compression problem? I hadn't a clue. But I asked the questions nevertheless.

"Did you grow up in a big city?" I asked.

"No, I was a farm girl," she admitted. "We had a lot of chores. Even though we had a big family I was always busy with something or other. My mom was sickly most of the time, and it seemed like I got asked to do the 'female' things. You know, things that needed a woman's touch, though there were precious few of those."

She mused and bit her lip. "Seems like it was never good enough, though."

I was intrigued. "What about brothers and sisters?"

"I didn't have no sisters—only brothers. A whole mess of 'em. There was ten boys, all spread out but spaced close together. Maybe that's what got my mom so sick. My dad, he relied on the boys a lot. It was a pretty big farm back in West Virginia, and he needed a lot of help. The boys did the heavier stuff, harvesting, fixing machinery, things like that. Daddy got so he couldn't run things without my brothers helping him."

She brooded again, and the omega sign appeared prominently on her forehead. "But me, now, that was something else. I was the youngest of the crowd, the afterthought. Seems like I was just in the way most of the time. I dearly loved my daddy. When I was real small, he used to make me feel so good by swinging me around and hugging me. But my dad loved having the boys. Especially as he got older, he praised what they did for him. They were his 'right arm,' he used to say. In the later years he didn't have much good to say about me. Manure flows downhill, and I was at the bottom of the ladder, sure enough." Her manner became agitated, and her voice trembled with emotion. "I remember so clearly how he used to bark his commands at me, pointing his finger like I was guilty of some great wrong. He would say 'now you do this' or 'you do that, and snap to it.'" As she passionately described this, she waved and shook her right hand at me with the thumb and index finger fully extended and at right angles to each other in the position of accusation, her chronically stiffened posture.

As she did this, her entire body froze. She stared intently, curiously, at her fingers as though her hand were a pistol that had just fired a bullet, and the index fingertip was still smoking. I don't believe she really understood at this moment the significance of what she had revealed. She knew something odd was going on, but she still didn't "get it."

I got it, though. Mrs. Stowburn certainly had some unresolved issues from her childhood. I would not presume to play amateur psychiatrist. But in her melancholy words I heard a story of primal conflict between love and hostility. Unable to resolve this inner conflict consciously, she converted it to somaticized, physical manifestations—paralysis of muscles in her right arm—also known as a hysterical conversion reaction. Some people get gagging, choking sensations and become incapacitated from a memory that gives them a "lump in their throat." After a lifetime of inability to put to rest the intense emotions regarding her family dynamics, Mrs. Stowburn had converted them to a physical reality she hoped a physician could "fix."

She smiled benevolently at me. "Land sakes, I didn't mean to go on so like that. You should have hushed me up."

"Not at all," I protested. "I'm the one who asked you the questions. I've given it a lot of thought, and now I'm pretty certain you don't need a surgical operation."

She smiled more broadly. "Oh, that's nice. But then what's going to make my hand work better? I do have a lot of problems with it, you know."

How was I going to burst the conspiracy of mechanical manipulation that is surgery? The widely held hope and belief that locked in the flesh are the answers to unseen mysteries. The conviction that opening the skin, dissecting the vital structures, "letting in the night air," confers magical properties to the tissues. As residents in training, operating in the wee hours, we saw some patients get well after negative explorations and jokingly attributed the improvement to the "night air." After all, "A chance to cut is a chance to cure," right?

How was I to tell this simple, aged lady the answer to her dilemma was locked inside her brain?

People don't like to be told by their doctors, "It's all in your head, Mrs. Jones." And well they shouldn't. Because too often this trite little phrase is used as a substitute for the truth. If the doctor were honest, he or she would say, "I don't know what's wrong with you, Mrs. Jones. I know something is wrong, or you wouldn't be here. But I don't know what it is." But it's easier to save face and bolster one's ego by shifting the burden of responsibility, the "blame" for the problem, to the patient. The words are used as a put-down, a way to get the patient out of the office.

Yet the truth is, *everything* is in your head. The taste of chocolate ice cream, the color purple, the smell of gardenias, a good orgasm, *everything*. So the goal is to acknowledge the problem and find a solution to it, using all available resources. Rather than pile responsibility on the afflicted person, let's use a team approach to get answers.

On the wall in my waiting room, prominently displayed, is my credo. I wrote it in 1977 when I entered private practice. I had an artist friend letter it in calligraphy, so it looks a bit fancy. It says this:

> *I am not a healer; I am a helper*
> *You and I are partners in your care*
> *I can point the way, but only you can follow it*
> *For you are responsible for your own welfare.*
> *Ask me questions and I will try to answer.*
> *I will share my knowledge with you.*
> *But I cannot answer the unanswerable*
> *Nor can I perform the impossible.*
> *Doctors are human beings—not gods.*

In the early years of my practice, in the '70s and '80s, few people took notice of or commented on these words. It is particularly gratifying to me now that quite a few of my patients read the credo and mention it and its validity when we first sit down to talk in the examination room. A sign of the times.

Mrs. Stowburn had never noticed it, but its message was about to become relevant for her. Because, with expert help, she would become the primary agent of her own treatment and the proactive architect of her recovery.

The chief tool in her therapy would be hypnosis. Even the word itself excites visions of mysterious workings of the mind, of luxuriant fantasies and supernormal manifestations. For Mrs. Stowburn, nothing romantic lay waiting for her. Simply access to the dark crypts of her psyche and recognition of the warring forces of duty and desire that lay buried, hidden, within her psyche. Tied up with this emotional baggage was the trigger converting right-arm normalcy to right-arm pathology. This "smoking gun," if she could find it, was the key to her cure.

But before she could begin, I had to talk her into it.

"Mrs. Stowburn, I think there's a way to solve your problem and get your arm working well again. But the approach is, shall we say, a bit unconventional." I described what little I knew about the subject and told her about the psychotherapist whom I would ask to do the work with her.

"Sonny, I'm an old lady having trouble with my arm. I can't take care of myself or do my day-to-day chores. What have I got to lose? Anyway, you're very nice and I trust you."

She beamed her sunniest smile at me, her forehead smoothed, and I knew the hard part for me was, mercifully, past. For her, it was just beginning.

Eight months later, she reappeared in my office. Her smile was warm and all-encompassing, infinitely reassuring. She held up her right hand proudly for me to see. Her thumb and index finger moved well, although the Heberden's nodes, the arthritic bumps in the index finger, were slightly larger. Some bruising was still present. She handed me a piece of freshly made dark bread.

"I thought you might like to taste some of this. I baked it this morning. I'm not happy with the yeast—I need to get a fresher batch."

"I'm honored to eat your creation," I said approvingly. "Your hand seems to be functioning well. Was it a worthwhile experience?"

She paused and fixed me with a penetrating stare, deep and brooding. The psychotherapist had sent me terse progress reports, but they focused on her physical progress, not her mental state. I almost wished I hadn't brought it up.

"I learned quite a bit. A lot of it involved my family. Things I never knew about; things I knew but forgot; things I know now and will wish for the rest of my life I didn't know. Worthwhile? I suppose. But it's like making a pact with the devil. You get something in the exchange, something you were sure you wanted. Then, after, you wonder if you made a bad bargain. But it's all me, isn't it? Sometimes we look in a mirror and we don't like what we see. It doesn't do any good to break the mirror. If you don't like the message, don't shoot the messenger. Change the script. I can't undo what's gone before. But I can try to be different from what I was. I just hope my fingers don't get all stiff again! . . . How do you like my bread?"

4.

Phantom Memory

Mickey Mouse is deformed.

But so, too, are Bugs Bunny, Fred Flintstone, and a host of other familiar cartoon characters.

And they're all deformed in precisely the same way. They have a thumb and only three fingers on each hand.

You probably were unaware of the deformity. If you didn't notice, it's a tribute to the skill of the animation artist, the dexterity of the character's hand, and the legacy of Walt Disney and his longtime friend and partner, Ub Iwerks. Iwerks was one of the fastest and most inventive animators who ever lived. Until his time, cartoon characters like Betty Boop were drawn with a thumb and four fingers. Very realistic, but time-consuming. When Iwerks invented the Mickey Mouse character and had to crank out thousands of animated cels to meet deadlines, he undoubtedly discovered that a balanced, smoothly functioning three-fingered hand passed muster and was unquestionably easier, quicker, and a lot cheaper to draw. The rest is history. We don't notice the absence of a cartoon character's finger because cartoon hands move and function normally. We don't count fingers, we note spaces. If there is a hand with amputation of a central digit, with a short stump creating a visible gap between normal fingers on each side, the loss of the finger grabs our attention as surely as a missing soldier in a drill formation.

When John Furman came to me for evaluation and treatment, he had lost the end two-thirds of his right ring finger in a punch press injury at work ten years earlier. Unable to compete in the industry as a machinist, he became a high school shop teacher. A generally good-natured, cheerful optimist, John found, to his constant embarrassment, that when he tried to grip nails, screws, or bolts, they would fall through the gap between his fingers. The sizable space between the small and long fingers was quite noticeable. Students would ask him what had happened to him, and he tired of explaining his handicap. He wanted a solution.

John had suggestions involving nonexistent technology right out of *Star Wars*. They relied on seamless interfacing between inert metal and living tissues, and elimination of scar—both biological impossibilities and completely unrealistic. When I pointed out his false optimism, he looked crestfallen. "But I'm missing a finger. You're telling me there's no way to get it back or have a new one. There must be *something* that can be done to replace it."

I explained to him there is a surgical procedure called metacarpal transfer, nicely tailored to this problem. It creates a smoothly tapered, three-fingered hand, eliminating the amputation stump entirely and creating a strong, balanced hand with normal motion. At close range, it works with fluid grace and passes cursory inspection without calling attention to the absence of a digit.

I described the proposed operation in detail. First I would amputate the ring finger, handling the nerves properly and burying them, but I wouldn't stop there. Making accurate measurements to ensure the digital lengths were precisely correct, I'd use a microsaw to cut across the base of the small finger metacarpal bone. Leaving all the tendons, nerves, and blood vessels intact, I would then lever over the entire small finger onto the remnant of the base of the ring metacarpal, using smooth stainless steel pins to hold it in position.

"Rotation of the transferred digit has to be carefully controlled," I explained, "so the fingers can't scissor or cross against each other when making a fist. That's where the stainless steel pins come in. They're perfect for positioning the bones." John was nearly hypnotized.

The web spaces between the fingers are an important cosmetic feature. I assured John I would work hard to be sure the newly created webs looked natural. And, of course, the flexor and extensor tendons, which close and open the fist, had to be protected, intact and uninjured so that finger motion could begin right away.

A blank, somewhat worried expression appeared on John's face. If he was a bit bewildered, it was understandable. I wondered if I was pushing him along too fast with all the technical fancy footwork. I sensed he needed the explanation for reassurance, yet even for a mechanically adept machinist, there were limits. I backed off.

"Of course, many of the details simply add finesse, once the basic work has been accomplished."

"Of course," John agreed, knowingly, like a colleague in close collaboration. He was getting into the spirit of our "game," which would become serious if we jointly made a commitment to proceed.

John took a deep breath. Overwhelmed by the scope of the proposal, its complexity, and its sheer audacity and ramifications, he seemed speechless. But he also seemed a little furtive. There were other reasons for his reticence.

"There's something I haven't told you about yet," John said, "or anyone else, because it's too crazy. Even though I know it's gone, I feel like the finger's still there, pain and all. And it feels like it's still attached, still part of me. I still feel the ring I used to wear on that finger. Once in a while, I try to scratch my nose with the finger. When I'm absorbed in teaching a class, I even try to hold a pencil with it. I swear the feeling is real. Am I losing it?"

"No, you're not losing it, John. Let me see your right hand." My earlier examination had revealed distinct painful sensitivity over the cut digital nerves of the ring finger amputation stump. Where these nerves, smaller than a country matchstick, had been violently torn they had attempted to regenerate, a fundamental property of peripheral nerves. The combination of regenerating nerve and scar tissue had formed a small, exquisitely tender mass called a neuroma— literally, "nerve lump"—on each side of his stump. This was the basis for his long-standing pain. But why was the remarkably vivid image, the sensory impression of his finger, still there? The idea of it clearly spooked him.

The spectral force John experienced is the curious phenomenon known as phantom limb syndrome. Though its manifestations have been known to observers at least since the time of Galen, who wrote about it in the second century, for centuries physicians viewed the phenomenon as a manifestation of paranatural forces, almost beyond the ken of respectable scientific inquiry.

Only in recent times has "phantom limb" come to be seen as a normal and inevitable by-product of amputation. No one had ever explained this to John, so his reluctance to talk about it, his feeling that he was crazy, was understandable. For most patients, the phantom appears immediately, often so vividly they have difficulty realizing the amputation has already been carried out.

With time and healing, painful episodes subside in both upper and lower extremities, though affected muscles in the stump might occasionally jump or twitch, a phenomenon termed *jactitation*. Placing the stump in warm water will often relax clawed phantom fingers and promote comfort.

Commonly, on the application of an artificial limb, a phantom may lengthen and actually become identified with the prosthesis, with the lost limb perceived in the position of the prosthesis.

Oddly, phantoms may become more vivid or intensify after seemingly trivial reflex actions like yawning, coughing or even by washing the same side of the face with cold water. I asked John if such an unusual event had ever occurred.

"I don't talk about this—people would think I was nuts. But I don't go to horror movies anymore. I used to love them as a kid. Because, when I'm frightened, I feel the finger bending and straightening. I want to run it through my hair and feel it on my scalp. If you ever talk to anyone about this, I'll deny it."

His other complaints were more classic manifestations of the phantom at work. On occasion, the phantom finger would itch with maddening intensity. A sudden blow to the amputation stump made the phantom feel cold. Without any stimulus, a pleasant, warm tingling sensation would envelop the phantom for no apparent reason.

In the majority of cases the phantom is mild, fading somewhat over one or two years. Some researchers feel a phantom will ultimately disappear with no residual trace. Others believe this occurs only rarely. In about one-third to one-half of cases the phantom persists, even after ten years or more. John fell into this category. Several explanations account for the persistence of phantoms. If the original trauma was severe, the body seems to have an innate need to preserve some remnant of the part that was lost. A hallucinated remnant—a phantom—is, seemingly, better than nothing at all. If central pain persists, maintenance of the phantom seems to have a soothing effect.

My impression from John's bewildered expression was that he had had enough for one day. "I realize this is a lot to absorb all at once," I said. "And I'm sure you'll have lots of questions. So why don't we take a time-out to assimilate what we've outlined, discuss it with family, and come back together soon to talk some more. Is that okay?"

John again nodded his head, but this time absently. His expression told me his mind was elsewhere. The notion of a permanent fix was a siren call he could not easily ignore.

I saw him again within a week. His son-in-law was a practical man who, with the entire family, had discussed the procedure and its ramifications. John was obviously enthusiastic but he spoke a mile a minute in a tremulous voice. Correctly sensing his anxiety, they were cautious. Their spokesperson was John's daughter, Phyllis, who accompanied him to my office.

"So what would you do if you were him?" She stuck out her jaw, and I wasn't sure if it was to intimidate me or a personal quirk.

"I'm often asked this question," I replied. "I've never had a finger amputated, so I don't really know, but there is a good surgical resolution to his problem." I went on to describe to her what was, by now, becoming repetitious. All I could do was what I'd already done—try to explain to John, in understandable language, what the operation was all about and what his options were. My goal was to educate, I added, commenting that the word *doctor* comes from the Latin *docere,* which means "to teach." If John clearly understood all the ramifications of what I'd told him, then I'd be happy. Now it was up to him if he wanted surgery.

"Hrummph." Phyllis, like a medieval knight, had been charged with carrying the family's banner—and their skepticism. She was not about to give in so easily. "We have some reservations. How much experience have you had with this procedure? What are your results like? Do you have any photographs?"

I refused to be goaded into being defensive. "I've presented this entire procedure, soup to nuts, to hand surgeons at a national medical meeting. The patient I presented was operated on by my teacher. I personally have done this only one other time. I have photographs, which I would be happy to share with you."

Phyllis and John wanted very much to see those. Phyllis seemed satisfied, although her attitude was noncommittal. "It seems like you have the qualifications. . . ."

John took up the gauntlet. "What Phyllis means is that we trust you to do your best. I understand the operation and want you to go ahead with it. As soon as you can get it on your schedule."

I faced them both squarely. "Let's be clear about this. All operations carry some risks, and I've explained these in detail to John. I would not suggest the surgery if I didn't believe the potential benefits outweighed the risks. You want me to proceed?"

There is an old expression: better the devil you know than the devil you don't. Was I about to create a new devil to plague John? Certainly, there would be some pain and months of infirmity. What then? Would the end, for him, justify the means? Would he like it? Would his hopes be realized, or be dashed in a futile gesture of wish fulfillment?

John Furman represents one end of the amputation spectrum. The other end is murkier, darker, more sinister because of its severity and life-altering implications. John's problem was actually relatively straightforward and uncomplicated. Many patients I've treated had amputations considerably more extensive and serious than his.

Especially shocking was the case of a forty-five-year old man,

Eduardo Garcia, whom I was called in to see in 1982. His right hand had been badly crushed and mangled in a cotton gin at work. It was a bloody mess. Cotton fibers and soil were ground into every recess, around the metacarpals, deep into the palmar space. Although there were no fractures, bone surfaces were deeply abraded. Delicate intrinsic muscles within the hand were peppered with plant particles, the nerves covered with debris. Grease from the machine's gears densely coated the tissues, making his wrecked hand look like a bird engulfed in black tarry goop after an oil spill.

Despite aggressive surgical debridement (literally, trimming and "unbridling" the wound of debris and devitalized tissue) and massive antibiotics, his dirty wound became infected. The hand died a horribly grotesque, suppurative death and had to be amputated.

His postoperative course was punctuated by intense surgical agony requiring stunning doses of narcotics. His forearm stump, left open at first to minimize the risk of invasive infection, eventually healed. I had him fitted promptly with a well-constructed prosthesis.

But he had great difficulty functioning smoothly with only his nondominant left hand to service all his needs. He could not write, pay bills, sign his name easily. Shaking hands was awkward. Personal hygiene and dressing were hours-long ordeals. He could not make love as before. He could not do much of anything as he had before he lost his hand.

To add to his misery, he developed a troublesome phantom, which plagued him mercilessly. If an object loomed menacingly in front of him, he would often throw up his right arm to ward off or block the threat with his nonexistent hand. The spasmodic reflex jactitations of his stump were often violent and painful, even though he had the benefit of immediate manufacture of a prosthesis and psychological support and counseling. In World War II this treatment allowed American soldiers to adapt well to their amputations and manifest a low incidence of symptomatic phantoms. By contrast, one in ten French soldiers, who did not have such treatment made available to them, returning to broken, invaded homes, poor support, and no job prospects, developed painful phantom limbs in their amputation stumps. This was strong evidence for the important role of psychological factors in phantom limb symptomatology.

Ed's phantom hand often projected at odd angles to his forearm rather than remaining in line with it. At times it appeared as a vivid, dramatic presence that failed to obey the traditional, conservative characteristics of phantoms. Like obstacle shunning. Despite the sure knowledge that one cannot stick one's arm into and through a wall, that two objects cannot occupy the same place at the same time, his phantom hand would, with disturbing regularity and without effort, penetrate a wall or door and remain there. This was most disquieting.

Moreover, agony in his amputation stump continued to frustrate all rehabilitation efforts, despite skin grafts over the end of the stump and rapid fitting with a prosthetic hook. The forearm nerves were properly buried, and neuroma pain was not the issue. The pain was of a particular kind, caused by a malady called reflex sympathetic dystrophy, or RSD. Its occurrence is a clue to the mysterious, hidden functions of the hand. Understanding it requires some explanation.

We all know about the voluntary part of the nervous system. Pour a glass of water and drink it; hammer a nail; towel off after showering. There are myriad activities we perform effortlessly, without thought, yet they require both volition and an enormously complex set of motor skills. One part of the nervous system, the voluntary part, accomplishes these.

But as you were reading this, you blinked your eyes ten or fifteen times in the last minute without thinking about it. Unaware, you breathed ten times as well. Sure, you can override this automaticity, voluntarily breathe faster or slower, blink at whatever rate you want. But, if I distract you for a little while, you'll go back to "cruise control" and carry out these essential functions automatically once again, without conscious thought.

If I hold a stick of gum under your nose, you'll salivate automatically in anticipation of chewing the flavorful wad of inedible paste. If you're hungry, your stomach may rumble. You may sneeze. Or burp.

All these and many other bodily functions are carried out by a part of the nervous system called the involuntary, automatic, or autonomic, nervous system. It's divided into two parts, depending on which chemical neurotransmitter powers the action. If it's acetylcholine, the autonomic function is called cholinergic. If it's noradrenaline, adrenergic.

The cholinergic, or "parasympathetic," system is vegetative, sedentary, meditative. Ever watch someone faint? Not infrequently, I had to give a local anesthetic injection in preparation for a cortisone shot. It was a sobering reflection on the unbidden potency of the brain to watch a strapping, macho six-footer react in an unexpected way. First his eyes would widen as he saw the (actually rather tiny) syringe and needle. Then his pulse would slow and beads of perspiration would appear on his upper lip. The color would drain from his face, he would turn white as a sheet, yawn—and pass out. All this represented a massive outpouring of acetylcholine at susceptible motor end plates, the muscles reacting according to a designated program. The reaction is both overpowering and unstoppable.

The adrenergic, or "sympathetic," system is the opposite, the "go-go" part of the autonomic nervous system. An ancient self-preservation mechanism, the fight-or-flight reaction, is responsible. Under the powerful influence of noradrenaline (norepinephrine), the pupils dilate; breathing, pulse, and heart contractions increase; and blood flow is diverted to muscles and away from the digestive tract and skin, all in preparation either to confront a threat or run away. Of course, times have changed and there are no longer any dangerous animals to encounter on a dark trail in the forest at night. But the protective mechanism remains ingrained in all of us.

Injury, especially to the extremities, activates the adrenergic, or sympathetic, reaction. In most instances, the normal sequence of events allows the norepinephrine reaction to run its course and subside, leaving us a bit shaken and fatigued yet, otherwise, none the worse for wear.

But in a perverse, almost capricious twist of mischance the adrenergic system sometimes fails to switch off after an injury. To this day, despite enormous volumes of research, no one knows why. Persistence of the sympathetic response to trauma has a unique signature, a set of symptoms and complaints recognized by S. Weir Mitchell during the American Civil War and termed, by him, "causalgia." Pain is the primary symptom, a characteristic burning pain, severe and relentless, out of proportion to the degree of injury, so pervasive that even blowing on the affected arm and moving its hair sends the patient into paroxysms of agony. In the early stages, the skin may be warm and red. Later, the skin is cold, sweats abnormally, and becomes mottled. The joints become inflamed, swollen, and stiff, and the bones become brittle and break easily. Unfortunately, if it occurs, such sympathetically mediated pain is separate from phantom pain; it may coexist with it and does not respond to the same treatment.

Ed's problems with pain were long-lasting and difficult to help. Ultimately, a series of nerve blocks combined with a DCS—Dorsal Column Stimulation, blocking pain with a device that electrically stimulates the spinal cord—proved effective, though just barely.

Ed survived an overwhelming barrage of counseling sessions, aimed at preparing him for a new lifestyle. But his wife could not adjust to all the changes and, amid bitter recriminations, they divorced acrimoniously. Thankfully, later remarried, he slowly put his life back together and went on.

Lacking intellectual substitutes, his ability to do manual work had always been his major source of gratification and identification. Without his dominant right hand he became useless in the eyes of his coworkers. Work in the cotton fields was his only skill. He had to go to great effort to prove he could still do his job, to prove himself to others and to himself. Throughout his rehabilitation he maintained a richly expressive sense of good humor and, ultimately, lost the phantom and adapted to the use of a prosthetic hook, guaranteed *not* to be esthetic or romantic in bed. Even after great adjustment, the loss of his hand had a profound effect on him.

The story I've summarized in a few short paragraphs consumed fifteen years of this man's life. That his life and its subtleties revolved around his lost hand was no surprise. Before it was lost, I'm certain he never gave it a second thought.

Therein lies a crucial feature of hand function. We typically never give it a second thought. Using our hands is as natural as breathing. We ignore them and the myriad things they do for us—unless they are gone or don't work.

The journey of the hand is a profound journey, as profound as that of the mind since the two are inseparable. The hand is the working extension of the brain, the body part that converts fantasy into reality. Mind and hand are holographic. Both are capable of almost infinite subtlety. What the mind conceives, the hand fabricates. What the mind wills, the hand carries out. Good or evil, since the hand is not judgmental. It is the willing servant—supremely competent—but, nevertheless, humanity's most revered and most indispensable tool.

John Furman decided to carry out the surgical operation that would rid him of his troublesome amputation stump. He was even more talkative than usual in the operating room. The anesthetic block used to put his arm to sleep was completely effective. Tucked in like a bear in a cave under a formidable barrier of sterile drape sheets, he quickly exhausted conversation with the anesthesiologist. Bored, he became intolerably chatty and I had to beg him to quiet down so I could concentrate on myriad technical details. Like many bright patients in similar circumstances, he wanted a blow-by-blow description of exactly what I was doing. He bemoaned the absence of ceiling mirrors, which, he assumed erroneously, would allow him to see my work, the way you can see what a dentist is doing to your teeth from the reflection in his eyeglasses. John became silent only when I began to cut the metacarpal bones with a microsaw. The sound is jarring. Dental drills, moving at much greater speed, are higher pitched with hardly a change when they grind teeth. But when the saw cuts through bone it makes a rasping noise of resistance unmistakable even to the uninitiated. The machinist in John must have shuddered. Perhaps the unlubricated sound brought back memories of the original amputation. He remained silent until I was reconstructing the webs as part of the skin closure. I remarked we were nearing completion and things looked fine.

He again became talkative, rambling with no focus. He talked about the machine shop, the OSHA violations that had led to his accident, the shock of seeing his finger dangling, his unfortunate lack of disability insurance. He talked about cartoon characters like Bugs Bunny, Tweety and Sylvester, and Elmer Fudd. Now, he said, his students would really have something to laugh at.

Six months after his fancy procedure I had to bring John back to the operating room for a minor touch-up, a five-minute revision of part of the skin closure, which hadn't healed exactly as I wanted it to. The phantom was gone and showed no sign of returning. John lay on the operating table, his arm outstretched on the arm board, with me seated beside him injecting local anesthetic. My scrub nurse, a scant eighteen inches from his reconstructed hand, commented, "Dr. Arem, what was wrong with this fellow's hand, anyway? It looks normal to me—I don't see any problem." My eyes met John's and there was a twinkle in his.

As I was applying the bandage, I asked John if he thought the whole adventure was worthwhile.

"Yes and no," he said. "Having surgery is a nuisance. But I got what I wanted. On balance, it was worth it. But now that the phantom is gone for good, I kind of miss it. Even a phantom finger is better than none at all, I guess. My shop students have given me a new nickname: 'Mr. Toon.' But when I get away from school, no one even gives my hand a second look."

He was thoughtful for a moment. Then he said, "Walt would be proud of both of us, wouldn't he?" And, with a flourish and great emphasis, he added, "At least, he would admire the artistry."

John's good nature and sense of humor were his best allies in recruiting sympathetic help for his quest. They had gotten him this far. The rest was in the lap of the gods who, I prayed for John's sake, were equally good-spirited.

5.

Camouflage

Sam and Maggie were tight.

By that, I mean they were a dedicated couple with a committed relationship. Inseparable. I'm sure they would have entered my office hand in hand were it not for the bulky bandage on Sam's right hand.

Of course, *tight* also has a connotation regarding alcohol consumption. Maggie told me what the cockney bartender at the local bowling alley once said to her: "Margaret"—being a gentleman, he always referred to the ladies by their proper first names, Samantha and Margaret—"you always pays your bills, and you pays in cash. No IOU's to worry about with you two. What you do when you're on the road is no concern of mine. Here, your credit is good with me." With that, he refilled her scotch and soda.

Samantha Dayton indeed looked very proper, sitting in my exam room wearing a beige business suit and low heels. Maggie had come more as herself, slightly tipsy, in jeans and flats. Sam fidgeted, but anxiety didn't show in her raspy voice.

"Doc, I've been told you're the one I need to see. So here I am. Fix it." She waved her bandaged hand in front of me, almost belligerently. She absently picked her nose with her left index finger, then didn't know where to put it. I could tell by her aborted gesture her instinct was to wipe her slimy finger on her clothes, but the business suit was the wrong receptacle. She spotted a box of tissues on my exam table and leaned toward it. I caught her panic and silently pushed the box over to her. She appeared relieved not to have to ask.

"First, it would seem appropriate to find out what's wrong. Before I remove the bandage, why don't you tell me what happened?"

Sam inspected her left hand. Her guilty finger was now clean. But nothing could hide the calluses, rough skin, nicotine stains, and uneven, broken fingernails. "Well, I was cookin', you see. I don't do that a lot. Maggie and me, we eat out, mostly. But we were in a big hurry, and I was fryin' up this slab of meat. Somehow, I don't know how, but the grease in the fry pan caught on fire. I tried to blow it out, but nothin' worked. So I moved the thing across the room to the sink. As I went to throw the whole mess in the disposal, a bunch of the flamin' grease splashed onto the back of my right hand. Oh my, let me tell you, that smarted. It hurt like shit, is what it did. It kept burnin' and I couldn't get it to go out. I screamed and screamed, jumpin' around like a crazy woman." She gesticulated wildly, bouncing up and down in her seat. "I even knocked one of our bowlin' trophies off a shelf onto the floor. Lucky it didn't break. Maggie heard me and helped me wrap a towel around my poor hand." Maggie grinned like a Cheshire cat and blinked her eyes.

"That squashed the flames, but my hand was a mess—all blistered an' peelin' an' icky. But I didn't feel so much anymore, that was the good thing."

Not hurting wasn't so good, but she didn't know it wasn't, or why. Normal skin has feeling transmitted to the higher centers by living nerves. As long as pain persisted it meant the nerves were still alive, enough to send a pain message to the brain. In a burn, the absence of pain is ominous. Such a blissful reprieve may mean the burn is deep enough to have destroyed the nerves, converting a second degree to a far more serious third-degree status.

"What did you do then?" I asked.

"Maggie an' me, we hustled over to the hospital emergency room. I almost wished we hadn't. What a zoo! We had to wait for almost two hours. There was kids pukin' their guts out, all kinds of people with bloody bandages, pregnant teenagers ready to have their babies. One of 'em had dropped the kid in the car on the way there.

More carryin' on, I swear. But I wasn't much better," she said, waving her arms. "After a while. I was bitchin' and swearin' an' ready to walk. God, I wished I had a drink, but right then the nurse came an' led me to a room, where we had to wait another friggin' half hour. The smell made me nauseous." She grabbed her neck in a mock throttling motion.

"Finally, some geek with thick glasses and a white coat came in an' announced he was the doctor. Well, yip-pee. I really wanted to vomit now, but Maggie put an elbow in my ribs an' made me shut up." She winked at Maggie.

"It took him a little while to decide what to do. He got on the phone with some other dude an' nodded his head a lot, like he was gettin' good advice. Then he came back, smeared this white pasty crap all over my hand an' put on a bandage. He gave me a name and address and said to go see this guy, soon. An' that was it. I went outside to have a smoke and thought about it. No way was I goin' to go to this guy, who I guess was too busy to come see me in the hospital. So I asked around and got some other names. An' here I am."

"I have to take off your bandage to see what we're dealing with. You're not going to like me for a few minutes. I have to hope the ER doc put some nonstick barrier between your burn and the gauze."

"Go for it. But first, I'd like to go outside an' have a smoke, if it's OK. I'm a wuss when it comes to this sort of thing. Do you have time?"

"Go for it. I'm not happy about your cigarettes, but we'll deal with that later."

I opened the sliding glass door to the back patio to let them out. Sam kept pulling her suit down and opened her shirt button. Her first drag was an especially long, slow one. As I watched them relax under the pineapple guava tree, I saw Sam shudder despite the warm sun. Between puffs, she pulled at her waistband with her left hand, fiddled with her stockings, and stretched her neck like a turtle whose shell is too restrictive.

Maggie was serious, attentive, and obviously more comfortable in loose clothing. Eight pens were stuffed in her shirt pocket, the remains of a pack of Camels bulging behind them. Sam looked unhappy, but there were a lot of potential reasons why. I hoped time would loosen her up.

Sam finished her smoke and threw the butt on the brick floor, grinding it viciously with her heel. Then she sauntered back in.

"Okay. I'm ready now." She smiled.

Armed with bandage scissors, I trimmed away the gauze. Gratified to see a layer of nonstick Adaptic covering her charred skin, I knew I could undress her burned flesh painlessly. It's amazing how far reduction of pain can go to engender confidence in a doctor.

I would need all the confidence from her I could get, because the burn didn't look good. The skin had blistered and peeled almost immediately, signifying, at minimum, a second-degree burn. First degree, like a simple sunburn, produces only redness. The presence of blisters meant at least part of the skin's thickness had been destroyed. The critical question was, How much?

I gently washed the wound with sterile saline to remove the white Sulfamylon, which had been troweled liberally onto the burn. This true miracle salve, an unpretentious antibiotic cream, has saved countless lives, preventing catastrophic infection in major burns. Having proven itself, it was now widely available and stocked in every emergency room. So anyone could—and did—get at it and use it, often indiscriminately. Although Sam's burn was substantial in terms of the amount of skin needed to cover it properly, it seemed small when measured as a percentage of her entire body surface: it was localized and not a threat to her life. The antibiotic cream actually retarded the formation of a new skin surface, slowing her healing. But smearing it on the burn sure made everyone feel better, whether it was appropriate or not.

With the cream removed, I silently studied the swollen surface. Many of the smaller surface veins were clotted, but this had no real significance and didn't imply thrombosis of the more important vessels. The sensory exam was critical.

"I'm going to stroke your skin lightly with this sterile cotton swab. I'll move from place to place on the back of your hand. Sing out the instant you feel the touch." Tight-lipped, she nodded agreement.

"Okay, here goes." With my other hand, I held up a notepad to screen the swab's position from her curious eyes. After several false passes, I made the barest contact just beyond the wrist.

"Yo." She didn't hesitate for an instant. Good. Intact feeling meant survival, at least in one area. As I moved inch by inch, I was able to map the burn, which extended from the wrist across the entire back, or dorsum of the hand, to include the web spaces, the first segment of each finger and a short way onto the thumb. Some parts had no feeling at all and were dry and leathery. This observation meant curtains for skin so affected. A burn was an injury, and skin injured but still alive responded by swelling and weeping fluid. Dead skin was, quite literally, like the tanned cow skin you wear on your feet. Sam's burn was 50-50, part superficial and part deep.

"You ready to hear it?" She wasn't going to like what I had to say, and I wanted her with me, attuned to me, without distractions.

"Go for it, Doc." She lowered her skirt zipper and pulled Maggie close to her, squeezing her friend's hand tightly.

Maggie blushed and squeezed back, careful not to hurt Sam's remaining good fingers with a grip from her well-muscled forearm.

"What makes skin feel normal, what gives it suppleness and elasticity, is the dermis, the layer immediately below the surface. The surface layer, the epidermis, is the part that thickens and makes calluses in response to mechanical rubbing."

Sam studied her left hand as I spoke. It mirrored her habits. The bowling bumps, the smoking, her lifestyle. The scars, too, reflected history she told me about. Street fights; accidents with firearms; inexperience fixing an oil leak. Like most people, she took her hands, and what they did, for granted. Now, with her dominant right hand out of commission, she was both humbled by and furious at her helplessness.

"Partial thickness skin injuries heal by growing a new layer of epidermal cells over the surface. Glands and hair follicles extend from the surface deep into the dermis and are a source of these cells. When you were a kid and skinned your knees falling off your bike, you scattered your skin cells all over the pavement and made bloody scabs covering the abrasions. Wisely, your mom probably told you to leave the scabs alone until a new layer of pink baby-new skin had formed under the scabs."

Sam snorted, a short, derisive, exclamatory burst. "My mom was usually soused and didn't have a clue what I was doin'. Or care. Dad was long gone; I never did get close with my half sister. Whatever I figured out, I figured out on my own. It's always been like that. Until I met Maggie." She glanced lovingly at the plump, energetic woman sitting next to her, and squeezed her hand tighter.

"The new knee skin looked and felt normal because the abrasions were superficial and almost all of the dermis was preserved. With deeper injuries, though, more dermis is lost, more scar forms, and the quality of what's left is poorer. If the burned skin overlies mobile joints, the scar tissue, contracting and shortening as it forms, can pull the joints into a fixed, grotesque posture."

"So what are you tellin' me, Doc? I'm gonna be a gimp for life? I'm condemned to sell pencils on a street corner, or peddle newspapers an' never bowl again? Christ! If I became a vegetarian now, it's too late."

"What I'm saying is that it wouldn't be satisfactory to simply re-dress the burn and leave it alone. In my judgment, it would heal, but the functional quality of the healed result would be poor. Patchy areas of the burn are pretty deep. It would make a dense scar pulling your fingers back so tightly you'd be unable to grip or pinch anything. Much less hold a bowling ball. What's your average, anyway?"

"It's around 210, give or take. I'm team captain this year. At least, I was. I need a drink." She pulled a small silver flask out of her inside pocket and took a swig.

"So what choices have I got, if any?"

"Actually, there's one choice, and it's a good one. It involves surgery. I have to remove the entire burn, cut off all the burned skin and throw it away. Then I have to skin-graft the wound, using a large sheet of partial thickness skin from your behind. The donor site back there heals like the abrasion on your knee. But instead of the removed skin winding up in tiny fragments all over the pavement, I have it to use as one large piece. Technically, it's dead as soon as I cut it off your body. But if I put it onto your skinned hand, dress it carefully and do everything just right, it picks up a blood supply and comes back to life. Since it carries with it a reasonable amount of dermis, your hand should look fine and work pretty normally. What do you think?"

Sam eyed me like a caged animal, flexing and extending the fingers of her left hand. She put the flask away and glowered at me.

"Ain't there somethin' you can put on this? Some kind of artificial skin? I mean, these are modern times. There must be somethin' some guy invented to make the surgery unnecessary." Hopeful anticipation flashed in her eyes.

Actually, Sam was on the mark. Development of a skin substitute has been a biological dream for centuries. Particularly in large burns, in which there is not enough intact donor skin to resurface the skin which has been lost, a dressing that retained moisture, prevented loss of body heat and calories, and acted as a barrier against infection from ubiquitous bacteria would certainly be worthwhile.

"Sam, a skin substitute should, ideally, be adherent, control evaporative body fluid loss, be sterile without allergic, toxic, or infectious properties, and be durable, flexible, stable, easy to put on and remove, available and not prohibitively expensive."

She absorbed the information silently, hope fading. Nothing works as well as human skin in replacing lost skin. Sam was certainly justified in asking for an alternative solution. Its time had, simply, not yet come.

"What if I say no? What if I just walk out of here? Why not cut my losses and keep what I've got? This is messin' up all my plans." She threw up her arms, scowling.

Before I could speak, Maggie, who had been silent, grabbed Sam and spun her around. "You idiot. We're in this together. We had an agreement. I don't fink out on you, you don't fink out on me. Where am I gonna be if you can't hold up your end? Are you some Looney Tune, or somethin'?" She blinked her eyes and smiled that Cheshire cat smile again. "Anyway, hon," she murmured, pulling Sam closer, "I like what you do with two hands, not just one."

Sam opened her mouth to reply, but I interrupted. "On a

purely clinical note, Maggie's right. The window of opportunity is closing. If I don't remove the burned skin and put on a skin graft within the first three or four days, the process of wound contraction begins. The graft shrinks and tightens, lousing up future hand dexterity and softness."

Sam couldn't move, pinned by Maggie's embrace. Like Mr. Hyde reverting back to Dr. Jekyll, she lost her ferocity and became pliant. She gazed affectionately at Maggie. "Do it, Doc, before I change my mind. I'm a prisoner of love," she said, humming the song. "Anyway, our league doesn't start for a month or so. I'll be all fixed up an' ready by then—right?"

"Close," I said. My mind raced ahead. Set up the admission. Book the surgery as urgent. Rearrange the office schedule. Lock it in. There would be time to discuss the downside, the risks, the negatives. Did the positives outweigh the negatives? I wouldn't operate if I didn't think so. But though everyone wants a guarantee, I can't give one. Too many uncontrollable variables. The best I can do is to be truthful.

Nothing in her experience had prepared Sam for the next few days. Admittedly, she was no stranger to emergency rooms. A rough-and-tumble veteran of half a dozen brawls in and out of bars, she sported the scars of her misadventures.

With surgery just ahead, Sam needed to remain focused, stick with a reasonably healthy diet (eat, rather than drink, her calories), and be a model citizen, at least for a little while. She assured me she would. When she was healed, she could resume her nontraditional path.

Sam didn't need to spice up the stories about her. One look at her battle-weary face was convincing evidence she was the living embodiment of her creed. Maggie had confessed to me, in solemn tones, the bar's inhabitants knew and grudgingly respected her. Sam, I was sure, wasn't trying to impress.

Sam was her own self: blunt, feisty, crude. What made the current situation so hard for her was her sudden dependence on others. And she didn't like being a team player.

"Why do I have to stay in the hospital?" she asked in a belligerent tone during the admission physical.

"Because, for grafting purposes, the true size of the burn is enormous," I said. "I use a special instrument called a dermatome to remove sheets of skin from your butt. A setting on the machine lets me predetermine the precise thickness of the skin grafts. I then have to sew the grafts into place. I'll tailor the incisions along the sides of the fingers and webs to minimize visible scarring and maximize your function. But it's a lot of sewing and it's going to take a long time, at least three or four hours. I'll inject long-acting local anesthetic everywhere when I'm done, so you won't have any pain in your hand or your rear end when you wake up. But it's simply more than I can comfortably do under regional or local anesthetic. So we'll put you to sleep." As part of a complete physical exam I noted no evidence of liver enlargement, free fluid in her abdomen, or anything to suggest alcoholic cirrhosis. Her liver function studies were dead normal with a negative stool guaiac test. Guaiac is a dye that changes color in the presence of blood. A negative test meant that Sam wasn't bleeding internally. Despite her checkered and colorful history, I felt she could be given a general anesthetic safely.

Sam grumbled and gestured defiantly, but Maggie was there, as always, to reassure her, to soothe. She fluffed Sam's hair, even did a pedicure (the fingernails were beyond help). I took advantage of her calming ministrations to go over the details of surgery with Sam, and present the potential risks. The dermatome could malfunction. The skin grafts might not survive or "take" and have to be redone. The final scarring could, potentially, be less than optimal. But Sam was listening with only casual interest. For her, the hard part was showing up. Once having gotten this far, the rest, for her, was downhill. And, truthfully, I was pretty upbeat about the surgery and its aftermath. I didn't expect any problems.

When the time came the next morning to wheel Sam to the operating room, she was surprisingly chatty but pretty relaxed. Maggie, in sharp contrast, was a basket case. Her hair was disheveled, her complexion was ash gray and she paced the floor nervously.

I had come in early to provide reassurance by my presence. To look at Maggie, one would have thought the entourage heading out for Room One was a prison team conveying Sam to be executed. Maggie's cheeks were streaked with tears.

"Doc," she blubbered, "when am I gonna see Sam again? I've got a bad feeling about this whole thing."

"After surgery, she'll go to the recovery room until she's fully awake, about an hour or so. Then they'll call you in the waiting lounge and you can accompany her back to her room on the ward." Maggie clutched at Sam's bedsheets, sobbing.

I pried Maggie away and watched with some sadness her defeated struggle as she moved aside to let us pass.

Sam went to sleep peacefully, without a struggle. Her nonviolent acquiescence to the anesthetic surprised me. Even babies fight the mask and, given Sam's warlike reputation, I expected sparks, if not a conflagration. I positioned her on the operating table with her right arm extended out to the side on a padded arm board. When the time came to take her grafts I would prep and drape her buttocks and legs separately.

In the brilliant, unforgiving illumination of the twin OR lights, with Sam asleep and the tube to deliver the anesthetic gas secure in her trachea, I got my best look yet at the burn. Even better, scrubbed and dressed in a sterile gown and gloves, I could touch the skin painlessly and get a truer measure of burn depth. Although my professional persona was still on probation with this lady, I allowed my instincts to take a congratulatory bow. We were in the operating room doing the right thing for the right reasons. Most of the burn was as deep as I feared. I prayed that the final results, achieved largely through her efforts in therapy, would vindicate all the difficult decisions. I was especially pleased to note that the color match between buttock and hand skin was nearly perfect.

I made preparation to take the skin grafts, thankful we were doing so within only a few days after injury. The grafts would shrink a little anyway, even if they were put on five minutes after injury, but the unpredictable effects of scar contraction would be minimized.

Skin has elastic fibers that make it stretch. This property gives skin grafts remarkable handling characteristics. They are surprisingly tough, and toughness increases, as you might expect, with graft thickness. The outer surface has a fairly normal color, but the underside is white. When removed from the body, the graft curls inward dramatically, often rolling up into a tube. The curl makes it a cinch to lay the graft onto the recipient bed oriented correctly, raw side down. It doesn't work and the graft mummifies if you reverse it. But if you do everything right, blood vessels from the "bed" grow into the graft, which is destined to die without some form of nourishment, and bring it back to life.

The process is the same for all tissue grafts. Amazingly, thin skin grafts often take on a bright pinkish-red color within a few minutes after placing them on a good vascular bed. This color change, called inosculation, probably signals rapid early blood vessel ingrowth and augurs well for ultimate graft survival.

As I excised the burned skin, I was gratified to notice the large veins draining the back of the hand, still mostly soft and patent with liquid blood inside them. Those veins, in the normal hand, are a critical cosmetic feature. Without them, the hand looks pasty and unnatural, like a dummy in a wax museum. If you can pull it off, preserving them is an important form of surgical camouflage, especially if the final skin color match is close and the patient regains full wrist and finger motion.

And therein lies the secret. Slapping a skin graft on and getting it to take isn't that hard. The wounds heal, the patient gets some hand movement back and is happy to at least have something. The result is classified as "good," and everyone goes home satisfied.

But if your standard is normalcy?

So I preserved all the veins. The dermatome operated flawlessly and I obtained large sheets of split-thickness skin, which I sewed into position after stopping all bleeding with a nondestructive cautery. After putting on special bulky bandages to apply uniform pressure and reduce shearing forces, I checked the time clock. Four hours. Not too bad.

I changed, happy to get out of the damp surgical scrubs and into dry clothes. Working under scorching operating lights, I could appreciate the stamina of entertainers who, in tuxedos or tight dresses, also had to work for hours under hot lights. But I had no applause to reward me for my efforts. Just a quiet locker room and personal satisfaction.

I had one more important stop to make before I faced Maggie. Walking into the bustling recovery room, filled with the groans of patients reentering this world in pain, I found Sam stirring, just waking up from the anesthetic. Because most of my surgeries are done under local or regional block anesthesia, I rarely have the opportunity to take advantage of the twilight stage of recovery from a general, as Sam had been given. Patients in this half-awake, half-asleep state are hypnotically suggestible. Consciously, Sam didn't know I was there. I leaned close to her and whispered encouragement in her ear. I suggested she might find the pain was minimal, she was feeling well, and she would make a speedy recovery. Whether or not this helps, I figure it can't hurt.

Tired but optimistic, I went out to the lounge to find Maggie. She was dressed in a simple beige cotton outfit, sitting tensely on the edge of a frayed couch, staring out the window. She didn't see me at first, and I studied her anxious face. What kept this pair so tightly together? Maggie was like a prizefighter's cornerman, mopping Sam's brow, rubbing her muscles, soothing the hurt.

When Maggie's eyes strayed in my direction, she caught my smile and visibly slumped. Her weary expression betrayed strained relief. She flashed a questioning look.

"Yes—we're done, she's fine, everything went well, I'm pleased."

"I'm sorry if I seem like a pest. I swore I would behave myself. When my daughter was in intensive care, I practically pitched a tent here in the lounge. Some of the nurses still remember." She finished off her black coffee and crumpled the cup.

"It was a bad asthma attack when she was young. She's married now, living in Pensacola. But it was tough. My husband was a real shit. No help, abusive physically and verbally. Long gone and good riddance. If it wasn't for Sam, I'd have been down the tubes. She's a strong lady." She gazed longingly at the coffeepot.

"She'll be back in action soon. A couple of days, pain medicine if she needs it. Those skin graft donor sites can be sore."

"Sam's been through worse. So have I."

I had my answer. And I had affirmation of something I didn't have to be a trained plastic surgeon with three years in the lab doing wound-healing research to know. Not all scars are visible.

I hate, more than taking castor oil, to have patients admitted to the hospital. I hate the extra paperwork and the obligatory hospital rounds. More than anything, I dislike the habit of nurses telephoning at all hours of the night with minuscule questions.

My patients are usually not sick. Only their hands are sick. After surgery, if their pain can be controlled with pills, they can go home. Given a choice, people usually prefer to be nurtured by family and friends in comfortable, familiar surroundings. With experience, if you care about such things, you learn the tricks necessary to minimize postoperative pain and make this option realistic. But, once in a while, you can't do it. You might have to take bone grafts from the hip for a major wrist fusion. Or, as with Sam, you have to harvest large skin grafts from the buttock or thigh. The combination of a long general anesthetic and moderate postop pain usually mandates a one- or two-night stay.

But the first twenty-four hours had come and gone without a wrinkle. This night, the second after surgery, I expected peace and quiet. Betty McAddigan had the late shift. I've known her for years. "Blonde Betty" is a pert, wiry woman who burned out on emergency room and intensive care nursing and drifted back to the wards. The late shift and relative inactivity suited her temperament. One evening in the emergency room years ago, a sleazy young man from the inner city came in with acute appendicitis. The lice-infested hair from his pubic shave, sitting on a paper towel, began to move silently across the surface. Betty kept her cool. I trusted her.

So when Betty called, at 11 P.M., I was instantly on the alert.

"Dr. Arem, this is Betty McAddigan, with Ms. Dayton."

"Yes, Betty. I saw her just a few hours ago, and she was fine. I was planning to send her home tomorrow. Is her IV out yet?"

"Not yet. The janitorial crew keeps the hospital pretty clean. So when Ms. Dayton began plucking spiders off the wall—spiders that aren't there—I thought I'd better call you. No convulsions, no temp spike—just spiders. And she's a bit out of it."

"Keep the IV open. I'm on my way."

Any calm resignation or cautious optimism Maggie might have shown earlier was gone. To observe Sam, Maggie's heebie-jeebies were understandable. Sam was in a sweat, sitting bolt upright in bed, her eyes wide and staring, intently transfixed on the wall next to her. She was delirious, babbling incoherent nonsense about spiders multiplying, shaking violently, occasionally shouting at the bugs she alone could see. She picked at her bedsheet repetitively, oblivious to all of us watching her, including Maggie, who was beside herself with fright.

"What's going on? I've never seen her like this."

"That's because you've never seen her deprived of alcohol for more than a short while." I was most concerned about hyperpyrexia—elevated body temperature—but there had been no evidence of it. Betty, ever the pro, filled in some missing pieces.

"The delirium started suddenly, with no prodrome, no grand mal seizures, nothing. She has occasional lucid intervals lasting a few minutes, then lapses back into what you're seeing now."

Was this simple alcohol withdrawal, or was it the dreaded D.T.'s, delirium tremens which can carry a 15 percent mortality? I wasn't sure. It simply wasn't clear. Sam was delirious, yes. Tremulous, yes. But alcohol withdrawal can also produce all these features, then resolve smoothly with no ill effects. I wasn't taking any chances. I pulled Betty aside.

"Change her IV fluids to straight half-normal saline with one amp of potassium. No sugar—it's not a good idea for her to be given more carbohydrates. Stick some B vitamins in the IV as well. And, just in case it is D.T.'s, let's sedate her with five milligrams of Valium IV, very slowly. I'll write the orders." I pored over the chart with Betty, checking laboratory test results, intake and output.

When we glanced back, I was shocked to see Sam sitting calmly on the bed, looking a bit tired. Maggie was standing triumphantly at her side, a gleam in her eye. She held Sam's silver flask, which somehow had been smuggled into Sam's room hidden in the street clothes now hanging in her closet.

"You made it clear that all this weird behavior was withdrawal from alcohol. Well, I fixed that. One little drink never hurt Sam before."

Sam, now awake and coherent, smiled weakly. I nodded in grim acknowledgment. Although alcohol does in fact promptly end the withdrawal, it's generally not recommended for use in hospitals. Its low margin of safety is the major drawback. But it works, no question about it. Secretly, I wasn't displeased. There was simply no way Sam was going to stop drinking, no matter what I did or said. Maggie, responding to her own private terrors, saved me a lot of trouble and worry.

The next day Sam went home.

I saw her regularly afterward. Maggie brought her, stayed with her, took her away. Sam was uncharacteristically subdued. Was it an act for my benefit? I never found out for sure, but I think so. She had resumed bowling a few weeks after surgery, and her hand was improving steadily. She resumed drinking immediately. In fact, she never stopped. I think she wanted to placate me, so I wouldn't yell at her. She missed the point: I scowl disapprovingly, but I never yell.

At four months postop, the last trace of swelling had subsided and the skin grafts were fully mature, soft and pliable and their final color. Sam's efforts had paid off, and mobility of her wrist, thumb, and fingers was normal. She came in wearing Indian-made turquoise jewelry, a bracelet and rings to hide the scars where grafts joined original skin. Large, prominent veins coursed proudly over the back of her right hand. From a distance of fifteen inches, it was impossible to tell which hand had been burned.

"Well?" I said.

"Well," she said, "no more cooking at home. From now on, we eat out every night." She glanced at Maggie, who jutted her lower lip forward and nodded agreement.

"You're team captain. How are your bowling scores? Back up to snuff?"

"Haven't missed a beat. Better than ever." She gave me a vigorous thumbs-up.

"Maybe your concentration is better. Dealing with adversity forces you to think more about what you're doing. You can't just take it all for granted."

It was Maggie who responded. "We don't take anything for granted—least of all, each other." I remembered in vivid detail the forlorn expression on her face when Sam was wheeled away to the OR, and the terror of Sam's alcoholic delirium. Maggie was playing a dangerous game, with more uncertainties than usual, more than I would find acceptable. I wanted to reassure, but couldn't. Maggie had evidently weighed the odds and made a choice.

"Sam, do you remember anything about the spiders?"

"Funny you should bring that up, Doc. My mom ran an extermination service for a couple of years when I was a kid, to help make ends meet. To emphasize the kill potency of her chemicals, she bought secondhand military surplus clothes and equipment to make the outfit seem like we were engaged in guerrilla warfare. I've actually thought about doin' the same thing as a side business. The patterns are real pretty in camouflage. I don't know—it must be in my blood, in my veins."

6.

Plugged In

Lucy Alvarez fidgeted, uncomfortable in her starched whites, exuding exasperation, her breathing rapid and shallow. A pediatrics nurse for over thirty-five years, this diminutive, soft-spoken woman had befriended half the population of the barrio by taking good care of their kids. From the era of polio to the era of drive-by shootings and crack, hers was a steady, reliable presence the Hispanic population could count on. So what was she doing in my office? Good sense told me to be on my guard.

"I'm here—reluctantly, I must admit—because of your patient, Anna Soria. Reports received at the hospital suggest that her mother has been accused of playing an active role in producing Anna's burns. There are issues, not just of neglect but of abuse. Administration doesn't like it. My boss, Mr. Cornell, doesn't like it. Anna's becoming a hot potato. He's putting up red flags to block her surgery. Next, I fear he may try to ship her out."

I was irritated by the hospital's attitude. "Anna is a Mexican national. She came to Tucson for surgery only with the support of the St. Elizabeth free clinic."

"But the hospital still has to pick up part of the tab. My senior administrators claim they're concerned about issues of right and wrong," she said. "Catering to the mother doesn't look good. In addition, Anna is occupying a pediatric bed on a busy ward. Mr. Cornell says other children, American children, might be deprived of care because of Anna."

So that was it. Realizing I was picking my way through a minefield, I chose my words carefully. Lucy was especially attentive. "The story I got was corroborated by others. Frankly, I believe it. Anna comes from a small village deep in Mexico. They have some electricity, but very little else. Mama was cooking dinner on a hot plate, still plugged in, and walked away for a few minutes after removing the pot. The metal surface was glowing red, and perhaps that's what attracted Anna. She's only two, so we can't ask her. She put both hands palm down on the flat surface. She undoubtedly experienced immediate searing pain and screamed, but Mama was out of earshot and Anna was probably too scared to let go immediately. When Mrs. Soria found her, both palms were scorched. There's no medical care to speak of down there, so whoever attended to her put on some makeshift bandages and sent her home. What you see now is the product of spontaneous healing. It's unbelievable what the body does. Pretty horrendous, isn't it?"

Despite her training and experience, Lucy looked pale and shaken. As Anna's nurse, she'd seen the burned hands. But she wasn't finished yet. Having been put up to this by the hospital administration, she had a planned script and was not about to descend from her soapbox to engage in a meaningful dialogue. At least not yet. I braced myself for the verbal hurricane.

"Dr. Arem, Anna is completely helpless. She can't even grasp a small formula bottle. She's been committed to life as an invalid, in a culture which can't care for her. My bosses refuse to simply lie down and accept a passive role. They say this is a travesty of proper upbringing. They say child abuse can't be allowed to go unpunished." Lucy didn't mention potential threats to accreditation and funding for tenuous "free" programs. Surgery and hospitalization would be expensive and, with no disrespect to the St. Elizabeth clinic, who would pay?

I glanced at Lucy, who squirmed as if a lizard had run up her dress. She ran her thin fingers back through a tangle of black hair laced liberally with gray. Her eyes knitted in an expression conveying profound sorrow. We were old friends, and her expression revealed her pain at being forced into a position of being an administration toady.

"Lucy, let's be honest with each other. I believe Mrs. Soria's story. I suspect you do, too. Why don't we stop playing games and get Anna's hands fixed? We both know the hospital has other resources, other sources of funding. Cornell makes a lot of noise, but it's a cheap shot."

She cleared her throat.

"Dr. Arem, just between us, I've spent a lot of time with Mrs. Soria. She's incredibly distraught. As a single parent, she's worked harder than most to make ends meet and care for her kids, all of whom are quite young. It's cost her more than anyone may realize to get up here with Anna. But I don't think guilt is her sole driving force. She wants to do the best she can for Anna."

"I'm right with you, Lucy," I said.

She shuddered. "Those horrible burns need expertise if Anna's going to have any hand function at all. One look at them tells you so. There's no availability of such care in Mexico, even if she could afford it. So I think it's to her credit, bringing Anna here. Anyway, I agree. I can't believe she would intentionally do anything to harm her child." Lucy got up and paced the floor.

"I've talked at length with Mrs. Soria and it helped to not need a translator. Of all the kids, Anna's been real special to her. Looks most like Daddy, who was killed in a mining accident last year. With three brothers, Anna's the only girl. Mrs. Soria wanted her to go to school, to become an educated person. Now this. We all think we have problems. They pale by comparison." Lucy shook her head sadly. "Mrs. Soria and her sisters are counting on you to make it all right. They think you're a magician."

Of course they did. Everyone wants a magic wand. Only there isn't one. In Anna's case, I had my work cut out for me—there were some unique anatomic concerns.

I asked Lucy to accompany me on rounds later that day. Anna was sleeping when we arrived on the ward. Minnie Mouse and

Goofy, cavorting innocently, adorned the walls of her uncarpeted room. Bright red ribbons and a piñata hung from a ceiling air-conditioning duct. The mood was cheery but somehow formal. All hospitals feel the same. For a toddler used to living in squalor it was, I was sure, quite luxurious. We tiptoed silently to the bedside.

Anna slept the peaceful sleep of a much-loved child. A cushy brown-and-white teddy bear kept vigil next to her pajama-clad body. Lucy was shaking. She hissed with a sharp intake of breath. "Her hands," she whispered. "My God, look at her hands. No matter how many times I see this, I . . ."

When the hot plate burned Anna's open palms, it destroyed the skin as surely as cutting it away with a knife. With no other intervention, her body tried to heal the open defects the only way it could—by marshaling an efficient mechanism called wound contraction. Slowly, her fingers drifted together, curled inward and fused into a slag of scar and flesh. The net result was closed, healed wounds, but at a heavy price. In repose, or vainly struggling to open her hands, her fingers and thumbs remained tightly clenched. Only the very tips could wiggle slightly. Without treatment, they would remain this way for the rest of her life.

Lucy was unable to pull her eyes away from the sight of a lovely child with a hideous deformity. Her expressive face told me her professional instincts were at war with an insane reality.

"Lucy," I said, "when will I get the go-ahead for surgery? I'd like to take Anna to the operating room as soon as possible."

"Dr. Arem, since I saw you this morning I've already had it out with Mr. Cornell. He grumbled a lot—he's under pressure, too—but I think you'll have the approval within twenty-four hours."

"Want to join me in surgery?" I said. "I promise you an experience you won't forget."

"Yes. I mean, is it OK? Can you get permission?"

"Sure. We'll get your backup to cover your duties for a few hours. Shouldn't be a problem."

"I wouldn't miss it for anything."

"Let's schedule it for the day after tomorrow, Friday, at 7:30 A.M.

The OR's on the third floor, as you know." I admit my motives weren't purely educational. I wanted Lucy as an ally if the administration created any additional problems. They were an unknown, a wild card and, potentially, more trouble.

As we were leaving, I noticed Anna flexing and extending each finger to the extent of its mobility, rubbing her fingers over textures. A chill of recognition stopped me in my tracks. This was not merely mindless play, random and purposeless movement.

Anna was educating her hands.

We send our children to kindergarten at age five to begin their "formal" education. But the true educational process, the one crucial to our survival as a species, begins much earlier.

Scrubbing your hands in preparation for surgery requires concentration and attention at first. You're careful not to touch anything. You clean your nails scrupulously. You feel ridiculous, awkward, and self-conscious. But after a few thousand times, you get the hang of it. Like brushing your teeth, it becomes one of those mindless activities that gives you a chance to think about something else.

Scrubbing for Anna's surgery, I did a lot of thinking. I felt like a fighter about to do battle with a crafty opponent. Was that why my hands were shaking just a little? I'd settle down once under way. Having an audience added to my jitters. Lucy was in the room, eyes wide and nervous over the nosepiece of her mask, trying to stay out of the way as the scrub nurse set up the electric dermatome.

There was nothing routine about this surgery. I knew I'd be making it up as I went along. I felt confident I could cut away the scar and open the fingers. My biggest concern involved the nerves to Anna's fingertips. Her fingers had been tightly bent in a flexed position for months. While such a position would irretrievably wreck the joints of an adult, creating permanent stiffness, such joint stiffness was rare in young children. I knew that time and use would overcome most, if not all, of the joint changes. The real problem was shortening of the tissues surrounding the nerves running just under the skin. If the dense scar produced by the burn tightened the nerve sheaths too restrictively, I wouldn't be able to straighten the fingers without pulling the nerves apart—not a good thing for hand function. Until I physically cut away the contracted skin, I wouldn't know. The same tight scar would densely involve the arteries that ran next to the nerves to nourish the fingers. Pulling them apart might kill the fingers in a futile attempt to straighten them.

Starting on the right, with Anna's entire tiny arm prepped and draped sterile, I asked the anesthesiologist to inflate the pneumatic tourniquet. It was safe to operate for two hours in a bloodless field, but I was pretty sure excising the burned palm skin and scar from a hand this small would take a lot less time. Doing both hands at the same sitting, in sequence, would be unthinkable in an adult. Both hands would be bandaged and the individual would be helpless. But helplessness was no problem for a toddler, whose toilet and dietary needs were taken care of by nursing staff anyway. And to subject Anna to the pain of only one skin-grafting session was a kindness.

Wound coverage was a respectable trade-off for pain. Anna had lost her palmar skin, and skin grafts, properly harvested and sutured, would replace it. The idea of skin grafting (for nose reconstruction) was originally promoted by Indian surgeon Sushruta in the second century B.C. The idea did not catch on. Skin grafts were first used clinically in Paris in 1869 by Reverdin, who applied what he called "seeds of skin" to heal a forearm wound. Anna would need more than small seeds. Much more.

Over eight months had passed since the burn. The elapsed time allowed Anna's body to mature the scar, and I could establish better defined tissue planes than I could have if we had operated four months earlier. As I worked, I was gratified to release the thumb, which could be swung into a fully open position without tugging on the nerves or vessels. Score one for our side.

Lucy watched in silence. Her eyes still reflected anxiety, but I suspected fascination was replacing fear. I wanted to rope her in as I worked.

"Lucy, why do we have fingerprints? An amateur might say

it's so we can catch bank robbers. Amusing, but it's not the right answer."

Lucy had not expected to be spoken to. It was as if she were suspended in a bubble of unreality, watching the proceedings on a movie screen.

"I don't know," she stammered. "I never thought about it."

"Most people haven't. There are two reasons, really. If you look at the palms or the pads of the fingers with a magnifying glass on a warm day, you'll see tiny droplets of perspiration spaced evenly across the skin ridges. The ridges, which create the fingerprint pattern, act like the treads of a tire to generate friction. You need some moisture to make this work. Sweating is a product of intact, functioning nerves. As we age, our nerves don't work so well and our hands get drier—why we have to lick our fingers turning pages, for example."

Lucy nodded, wonderment in her eyes.

"The second reason is to increase surface area. The palm and fingertip skin ridges act like the folds of a radiator. It's hard to measure precisely, but if you unfolded the palm skin of an adult it would nearly cover the surface of a sheet of looseleaf paper. The hands are one of the major temperature regulators of the body. Attached to the arteries of the hand are small peripherals, to use a computer term, called the glomus apparatus. In response to nervous system control, these little gadgets cause the blood vessels to dilate or constrict, shunting blood on a massive scale in or out of the hands, like opening or closing major city streets to regulate the flow of traffic. It's why, on a cold day, your hands are like ice—you're diverting blood away from them to conserve body heat. The fastest way to get warm is to stick your hands in hot water. The glomi sense the heat source and open up the vessels. The hands turn bright red as blood rushes into them, absorbing the heat and carrying it back to the core so you feel warmer. Liquid works more efficiently than, say, a hot blanket or heating pad because it flows over the skin ridges and presents the heat to a larger surface area."

"So that's the explanation," said Lucy. "But what about Anna? The skin grafts from her buttocks won't have ridges. Will she be in trouble?"

"I don't think so," I replied. "The scar blood vessels that support and nourish the skin grafts have their own autonomous regulation. It'll take years to normalize."

As I continued to release the fingers the true magnitude of the burn became evident. It was horrible. What was left of the palm skin was a tiny nubbin of scar in the center of the hand. My conversation with Lucy hid my anxiety about the nerves. They were a little taut, but the fingers were straight and feeling would be preserved. I made measurements of the skin I would need, figuring about the same for the left palm. With the electric dermatome I harvested enough skin graft for both hands from the fleshy expanse of Anna's buttock. The extra skin would "keep" fine in a sterile dish until I got to the left side.

Early surgeons felt skin grafts, to work well, had to be thin. They became so thin by the time of the Franco-Prussian War as to be virtually useless, and the technique fell into disrepute until revived and perfected in the United States in the 1920s. The dermatome can be set to vary the thickness of the graft taken. For palm skin I wanted it on the thick side, so I set the machine for 18/1,000 of an inch. Buttock skin is over 80/1,000 of an inch thick. So I had excellent partial thickness skin to use to cover the burns. The buttock wound would heal like a deep abrasion, with new skin epidermis migrating from the edges and flowing up from the deep glands and hair follicles over the surface to heal it. The principle is one of the cornerstones of plastic surgery.

All of the hand dissection was done in a bloodless field with a tourniquet inflated. But there was no such luxury for taking the skin grafts, and the buttock bled a bit before I could inject long-acting local anesthetic to minimize Anna's postoperative pain and apply a bandage. As I painstakingly sewed the graft into position, I glanced over at Lucy, excited about how well the procedure had gone and future prospects for Anna.

My excitement melted away in an instant. Lucy was green. I'm not talking about a trick of lighting or reflection off the surgical scrubs. I have noted all kinds of skin hues, created by bile or blood pigments, and the range of colors produced by abnormal physiology in people is truly astonishing. But this was different from anything I had previously seen.

"Dr. Arem, I'd like to stay but I think I have to leave. Is there a problem with my going?" Her mouth was hidden by the mask, but it was probably open and I could hear her panting.

"Not at all. I'm sorry the surgery affected you this way. I don't like bleeding, either. It's probably why I gravitated to hand surgery."

"To be honest," said Lucy, "seeing Anna bleed made me ill. So I'll just let you fill me in on the rest. I'll be seeing Anna back on the ward. You'll let me know if there are any special instructions."

Lucy left in a hurry, her body language making it clear she was happy to get out. Surgery isn't for everyone. At least, I told myself, she had seen enough to know firsthand what I was up against. When Mrs. Soria pumped her for information about Anna, her analysis would, at least, have the ring of authenticity. I finished putting a bulky dressing on Anna's right hand, keeping the fingers and palm maximally extended and open. The graft would contract and tend to shrink, as they all do, but we would be starting off in the best possible position for her future function.

I really would have liked to quit and leave right then. But I was only halfway through and still had the left hand to complete, this time without an audience. I took a deep sigh and got back to work.

The burn on the left side was as bad as on the right. But I had "gone to school" working on the first hand and I knew the correct depth of dissection. So I could save some time. I was really moving along when I hit a snag.

I couldn't straighten the ring finger. The nerve on the little finger side of the ring finger was covered with thick scar and contracted. Dissecting and releasing the scar using my high-power loop magnification didn't help. The finger was bent down halfway, and there was no way I could fully extend it without pulling the nerve apart.

I sat there in a cold sweat and fiddled, trying to think and buy time.

My anesthesiologist, Ron Gates, looked at me over the top of the drapes.

"Well—what are you going to do?"

"I haven't decided yet. What do you think I should do?"

"Well, it's a nerve, after all. She's going to need feeling in her fingers. I think you have to save it."

"Even at the cost of being unable to grasp large objects? Of having a finger that's bent and perpetually in her way? She's an infant and her middle joint isn't stiff now, but, mark my words, leave it bent and it will be."

My gas passer was tired, and a bit ornery. He wanted me to be compliant, not confrontational. He had a bias and made what he thought was a reasonable suggestion. He growled at me.

But there were no right or wrong answers. We were down to the wire, and it was my responsibility to make a choice. A decision that she would have to live with—be stuck with—for the rest of her life. A decision I would have to live with, too. I felt like a spectator in an ancient Roman colosseum. Thumbs up, or thumbs down? Dare I choose? Those ancient Romans did, aided perhaps by a chalice of wine, but I wondered if they contemplated their actions the way I contemplated this one.

As delicately as I could, I took a deep breath and gently cut the nerve, allowing it to retract. The ring finger immediately straightened fully. The artery, adjacent to the nerve, tolerated the position without problems.

I couldn't see his mouth, but my six-foot three-inch colleague's eyebrows arched upward. Shock? Bewilderment? Anger? I couldn't tell. But it was clear he was unhappy.

"Wh-why did you do that?" His voice quavered. His bushy handlebar moustache was hidden by the mask, but I could imagine it twitching.

"Since it's just you and me, Ron, and since you'll comprehend what I have to say, I'll tell you. There are three reasons. There are probably more if we really understood the biology, but three's enough. In the first place, sensibility on that side of the ring finger is the least important for hand function. In fact, when feeling is missing from a critical area, like the tip of the thumb or index finger, hand surgeons will swing over a flap of skin, with the nerve attached, from the small finger side of the ring finger to restore feeling and make the thumb or index useful. It leaves one side of the ring finger completely without feeling, but there's an old plastic surgery expression: 'Never borrow from Peter to pay Paul unless Peter can afford it.' In this case, Peter can, and the absence of feeling in half of the ring finger will make no functional difference."

My colleague processed the new information silently. I had lost one audience and picked up a new one. Besides, talking helped me diffuse my own anxieties. I went on.

"The second reason is that the nervous system is immature and undifferentiated in infants Anna's age. When she grows up, whatever feeling she has in her ring finger will, to her, be normal, whether it truly is or not. She'll never even give it a second thought."

Silence.

"The third reason is that young kids have an extraordinary regenerative power. Anna's cut nerve will attempt to grow back together. In an adult, the large gap between the ends would be a major impediment to the quality of the result. But in an infant? All bets are off. She could possibly get back pretty decent feeling there. Given those reasons alone, the prospect of having a stiff finger perpetually in her way was unacceptable. I had to cut the nerve. Do you get it?"

It was done. Did it matter whether I received approval? I suppose looking for vindication was a way of spreading the responsibility to lessen the stress of it.

Neither of us spoke. There were no more surprises. I was able to fully straighten the thumb and fingers. What was left was, again, a huge raw area covering most of the palm (the true size of the burn that had destroyed the skin) leaving a tiny nubbin of contracted scar in its wake. For a second time, I had the tourniquet deflated, meticulously stopped all the bleeding with a nondestructive bipolar cautery, and laboriously sewed the graft into position, finishing up with a bulky dressing. Total operating time: about seven hours. No wonder my colleague and I were tired.

"Thanks for the good anesthesia, Ron. I appreciate your patience."

"Listen, Arnie," he said, "ignore me for playing the devil's advocate. We're in uncharted waters here. But you know a lot more about this stuff than I do. You have my vote of confidence. Will you give me a follow-up?"

We all handle the crushing weight of responsibility differently. One of the merciful aspects of anesthesia as a specialty was, I suppose, the fact that patient contact, patient obligation, ended in the postoperative recovery room.

"I have a better idea. Come and see Anna on the ward. Wait until next week, when I remove her bandages. She'll need a light general anesthetic to take the bandages off and remove her sutures, and you'll be doing it, I'm sure. I suspect there'll be some celebrating when she's able to use her hands again. Anna's brothers, aunts, and their kids have come up from Mexico, I understand. You won't want to miss the festivities."

"I'll try to make arrangements to be there." Ron wheeled Anna out to the recovery room, and I was alone. Soggy with sweat, fatigued beyond any quick rejuvenation, and alone. Alone with my thoughts, which immediately began to run the tape of the operation. Cognition is a two-edged sword, and instant replay is the surgeon's curse, a devilish reward for the ability to plan and execute. I craved sleep, but wondered if my mind would, perversely, invade my rest with frustrated "if only" longings. Nuances. I knew with a deep inner conviction that I had altered this child's fate. Only the years would tell her story fully. I checked Anna in recovery, then showered, went home, and plopped into bed.

Anna's family was jubilant. The bandages held her exposed fingers fully extended, totally reversing their previous functionless, clenched position, and the family could clearly see the change. In truth, even if her palms healed in this posture and her fingers had no movement at all, she would be better off than before. Of course, my aspirations for her were more grandiose.

Waiting a week for the grafts to "take," to become nourished by blood vessels growing into the grafts from the raw bed beneath, proved agonizingly difficult. I was as impatient as the rest, but I knew better than to rush things. We waited. Anna remained cheerful, preoccupied with two-year-old stuff, oblivious to the stirrings around her.

Finally, the day came. As soon as Ron had her asleep, I removed the bandages and sutures. It was hard to keep from jumping up and down with excitement. Both grafts had "taken" virtually 100 percent. Although I had saved a few small pieces of skin in antibiotic solution in the refrigerator, just in case, I didn't need them.

As soon as she was awake in the recovery room, Anna began to reach for objects and put them in her mouth. She didn't wait for or need permission. After all, she was a two-year-old. But this time, she had hands that were capable of doing her bidding.

By the time I changed and got down to the ward, Anna was enchanting the adoring onlookers who gathered to watch her. The piñatas had spilled their colorful contents on the floor and a bevy of children was there to scoop them up, squealing with delight. Minnie Mouse and Goofy, pulled irreverently off the wall, were being handed around like baseball trading cards. Nobody paid any attention to me. I moved next to Lucy.

She beamed. "Oh, Dr. Arem, isn't it wonderful?" I had to agree. Having completed the surgery, the technical part whose details were all but incomprehensible, I could put on a different hat. I could just be a person, and be enchanted like the rest. Ron came over.

"Arnie, you pulled a rabbit out of a hat and made a lot of people happy. Anna's life will sure be different now. You've argued many times that hands are what make us human beings. Watching Anna use hers I can really believe it."

Anna was playing with every loose toy in the room, picking up each in turn and putting it into her mouth. Like royalty, she calmly surveyed the throngs of well-wishers, her subjects, and seemed intensely satisfied. Of course, she regularly stole a glance at her mother, who remained quietly in the background. All manner of indulgence was permissible on this day.

I looked toward Lucy to ask her a question, and saw her

watching the door with her mouth agape, her thin body motionless. I followed her gaze and assumed the same expression. Coming into the room, carrying a large gift box brightly wrapped in red foil, was the hospital's senior administrator, Ralph Cornell. He spotted us, smiled tenuously, and came toward us.

I wasn't sure how I felt at that moment. I had corrected Anna's deformity but with no thanks to him. To me, this man was worse than simply passive aggressive. He had created barriers to my efforts. I took a dim view of anyone who made my life even more difficult than it already was. Before I could speak, Ron twirled his moustache and piped up.

"Well hello, Ralph," said Ron. "You told me on the phone you might be coming by. Your timing couldn't be better. Everyone's here. If you have something to say, say it now."

I watched, mesmerized, as Lucy called for silence and introduced her boss. His bearing and demeanor commanded respect, but Lucy must have weighed her options carefully before deciding to introduce him to this crowd. In any event, the immediate family knew who he was.

Ralph looked at the floor as he spoke. "Some of you know me; I feel like I owe Mrs. Soria an apology. This is a joyous moment, especially for Anna, who has her hands back. While I feel a great happiness for Anna, I'm like the bookie who bet on the wrong horse. I don't feel like I have any right to share in your pleasure. If I obstructed Anna's progress or was party to any delay in treatment, I'm sorry. I had some misconceptions, but my associate, nurse Alvarez, straightened me out, about Mrs. Soria, about everything."

Without hesitation, Mrs. Soria came forward and spoke, her voice soft and conciliatory. Lucy translated: "Thank you for coming here, Mr. Cornell. It was a very brave thing. What has happened is in the past. We must all live in the present and focus only on the future, because we will be there soon, and we want it to be pleasant and comfortable."

Wise woman, Mrs. Soria. Maybe Anna would go to college and become a great philosopher. She'd have a fine mentor.

Ralph beamed with the proffered forgiveness. "Thank you for accepting my well wishes. To cement my feelings of friendship, I brought a small gift for Anna." With a flourish and a quick flip of the wrist, he tore the foil and dumped the contents of the box on Anna's bed.

It was an enormous pink teddy bear, smiling a permanent cloth-and-plastic smile, beckoning hugs with its soft, thick, furry material and a great multicolored bow. Anna loved it immediately and threw her tiny arms around it, embracing it. A tiny bit of drainage oozing from the edge of one skin graft stained the fur.

"It's OK," said Ralph. "It's synthetic. It'll wash right off. Dr. Arem, I apologize. I was off base. Believe me when I say I'm truly delighted for Anna."

"Your apology is accepted," I said. "You don't make trouble for me, I don't make trouble for you. Jeez, I sound like a mafia don."

Ron reared up to his six-foot three-inch height, his thick moustache vibrating, and said in his best imitation Italian accent, "Hands off my friend, *capiche*?" We all laughed.

At that moment one of the family members gave Anna a formula bottle. She took it and plugged in, holding it in her mouth with her right hand. I had brought my camera down from the operating room and, unable to resist, took a photo. "This is what it's all about," said Lucy. "To be truthful, I'm much more relaxed seeing this than the operating room stuff." Like Kermit the Frog always sings, I thought with a silent chuckle, it's not easy being green.

Ralph comforted her. "I may run this place, but I don't like the sight of blood, either. I think it was gutsy for you to even go into the OR."

I was about to make some supremely wise, philosophical observation when one of the ward personnel came over and whispered into my ear. I nodded and faced my friends. "I have to go. They found me. The ER has an emergency for me. Some guy with a crushed hand. A giant teddy bear probably stepped on it."

Ron groaned and also began to leave. He was on call and, correctly, figured he'd have business.

Postscript: In 1994, fourteen years later, Anna showed up, unannounced, in my office. She was a lovely, black-haired sixteen-year-old who spoke only Spanish. My bilingual secretary confirmed she had come only with questions about the cosmetic appearance of the skin grafts. From my perspective, they were perfect. An amazing facet of skin-graft biology is that the grafts matched her growth. Even though, by actual measurement, they were four times larger than they were at age two, they were precisely proportional to her hands, which were also four times larger. Her fingers all had normal circulation and motion and were fully functional. Perhaps most significant, at least for me, was that the feeling in her left ring finger was, to her, completely normal. She was surprised I asked about it. Feeling was "normal" in both hands, and she never even thought about it.

Anna was an honor student and planned to go to college, fulfilling her mother's dream. She knew I had done her surgery but seemed unaware of the circumstances surrounding it. She happened to be in Tucson and didn't want to miss the opportunity to get her questions answered. I'm not sure what I expected. My excitement at seeing her was one-sided. She was a normal, healthy teenager who just wanted to get on with her life. People put behind them experiences that are unpleasant to remember. In Anna's case, she now had normal hand function and was too young when it happened to consciously remember the trauma. So her affect was neutral.

Mine wasn't. I felt relieved, vindicated, immensely gratified.

The truth is, surgeons thrive on such reinforcement. Teachers must feel the same thrill. The ability to touch a life and improve its quality. For me, it will always be a privilege.

7.

Circular Reasoning

Angela Bonfiglio didn't have a chip on her shoulder. She carried a log there, so weighty it was a miracle she could walk. Atlas, sagging with the world on his back, couldn't have stooped more than she.

She came to see me out of frustration, anger, and discomfort. Frustration because her hand problems robbed her of parties, many adult games, making home-cooked Italian meals—simple pleasures she came to take for granted. Anger because she blamed her work for causing it all. And discomfort, because she had lots of it.

"I need sleep. I'm so tired, I could spit. I can't function without sleep. Could you?"

I smiled a weary smile. Should I tell her about my surgical internship? Thirty-six hours on duty in the hospital, twelve hours off, for a year. Should I tell her about sleep deprivation so extreme my colleagues occasionally hallucinated and needed to be sent home? Should I recount fitful nights in an on-call room, being awakened every half hour by nurses for medication orders I couldn't remember the next morning? It was even more depressing than Angela's tale of woe. I decided against it, both for her benefit and mine.

"Every night it's the same damned thing. I fall asleep around midnight. At two A.M. pain in my left hand wakes me up. Not pain, really, not like at work. More of a burning numbness, mostly in my thumb, spreading up my arm. At first I was scared witless. I thought heart attack. I thought stroke. I thought brain hemorrhage. I don't know what I thought."

"You're too young," I interjected.

"Now, I know it isn't those—I'm still around—but it's uncomfortable as hell. So I stick my arm up in the air, dangle it over the side of the bed, get up and walk around, shake it, run it in warm water. I get my circulation back and, eventually, get into bed exhausted and fall asleep. For two hours. Then the same damned thing happens all over again."

Angela paced the floor, her hands tracing huge circles in the air. "I'm telling you, I can't go on like this. My transcription speed is falling off. I used to be one hundred forty words a minute error-free, and proud of it. Nowadays, I can catch and correct most mistakes on the new computer without leaving a trace. But if it was just a year ago, on my typewriter, I'd be using up correction fluid so fast the gals in the office would think I was guzzling the stuff. I'm the only one having these problems, as far as I know. What am I gonna do?"

Her story was classic and I knew what was wrong before I examined her. What surprised me most was the absence of similar stories among her coworkers. But it was 1991. How was I supposed to know I was seeing the early stages of an epidemic? The tip of an iceberg, growing larger, becoming all-consuming and pervasive, driving the medical marketplace.

"Tell me about work," I said. "You mentioned pain. What sort of pain?"

Angela stood up, her long brown hair unruly. "I've been an intake processor for DES for six years. I did the same work for the Sheriff's Department for three, right out of school. I spend most of the day interviewing applicants, writing out forms longhand in quadruplicate. It's tiring work. The state doesn't hand out its money casually." A kinesthetic person, she gesticulated wildly, outlining bigger circles in the air. "Then, in the last few hours of the day, I sit and crank out summaries. Now it's on a computer. At first, I was grateful for the new computer. I thought it was an improvement. Ha! What did I know? At this point, I'm not so sure."

"Go on."

"Well, 'round about two every day, my hands just give out. There's a sharp, nagging pain in my wrist, from the palm straight up my forearm. I shake my hands, but it doesn't go away." She demonstrated by shaking them vigorously. "I'm left-handed, and it's worse on the left side, but both hands hurt."

"At the beginning, would rest make it better?" I asked. "I mean, if you took a weekend off, or a small vacation, would it go away?"

"Yeah—at first. Funny how it would come and go. I could knead pasta dough with the best of 'em, and it would be OK. But after a while, the pain only went in one direction: worse." She jammed both thumbs downward. "Pretty soon, the pain wouldn't go away. It got to the point that writing checks, letters to my mother-in-law in Illinois . . . almost anything I did was excruciating."

"Your husband?"

"Mario's a production welder, works horrible hours. Kind of quiet, but he's no stranger to pain, I can tell you. Hot sparks burning his skin; heavy machinery tearing at him. He's not a complainer. He expects me not to be one. I don't get much sympathy."

"When did your numbness begin?"

"The numbness was funny, too. About a year ago. Both hands, but always worse on the left. I first noticed it driving my car. There's a series of tight turns I have to make to get home, and my fingers would go numb gripping the wheel."

"What about other times?"

"I hadn't thought much about it. But reading newspapers—hell, watching TV. It's supposed to make your brain sloshy and numb. That I can buy. But your hands? They'd go numb at the drop of a hat."

"What about the little fingers, the pinkies. Ever any numbness there?"

"Odd, now that you mention it." She absently stroked her small fingers. "Never, as I can recall. Even now, the thumbs are tingling but the little fingers are OK. Why? Is it important?"

"Very."

"Then you know what this is. You've seen it before. You know how to fix it?"

"I'm going to examine you first. Then, we'll see."

"Do I have to get undressed?"

"Uh, no. It won't be necessary." Wearing a smartly tailored outfit, Angela was seated facing me at a special desk I designed. She slid her chair close to the desk where she could support her elbows easily, while I examined her hands.

Sensibility first. But no pins. Many doctors test feeling with a pin. I don't. Nothing frightens a patient more than the sight of a sharp pin moving menacingly toward unprotected skin. It makes them guarded, withdrawn, and produces unreliable data. A moving, light touch gives more information with no fear of being hurt.

Beginning with the outside of the left thumb, I lightly stroked the skin to compare the quality of feeling with the outside of her small finger. "Which feels more normal?" I asked, moving from one finger to the other.

"The thumb is tingling and numb. I hardly feel it at all," Angela said. "The little finger feels normal."

"Good," I said, and moved to each side of the other fingers in turn. Gradually, a pattern emerged. The pattern culminated with the ring finger. In both hands, the ring finger was numb on the thumb side and had normal feeling on the little finger side.

Angela was astonished. "I never noticed any difference in feeling before. I didn't even know it was possible for feeling in a finger to be split. That's really weird. I'm sure there's some explanation. You're just being cagey." She pointed and waved an index finger at me. "Is this what you expected?" She winked at me, hope in her eyes, nodding as if she wanted me to do the same. To send a signal affirming her wishes.

"Yes," I said. "A sensory split in the ring finger is what I expected. I promise I'll explain all this anatomic gobbledygook as soon as my exam is finished."

Angela seemed reassured and allowed me to complete my evaluation. I found no surprises. Probably just as well. In my judgment, her problem was work related. I knew this would not go over big with her employer, who would smell increased workers' compensation insurance premiums in the wind, and, like many government bureaucracies, might be difficult to deal with.

The rest went quickly. Thumb muscles a bit mushy and weaker than they should have been in a healthy thirty-year-old. Tenderness at the entrance of the flexor tendon sheaths of the thumbs. Pain and tenderness over the muscle bellies of the finger flexors in the forearms when the wrists and fingers were passively stretched back. Nothing else. But those were enough to confirm what I already knew.

Now it was time to let it all hang out for Angela, to explain complex problems in terms she could understand. At least, I would try. She stared at me, and I thought I could detect both fear and hope in her eyes. And anger, held in check.

"The problem began with your work," I said. "Writing nearly all day, every day. Pressing hard to get through all those carbons. It's an impressive athletic performance, worthy of the major leagues—only it doesn't pay as well."

"No shit."

"When you write, applying pressure with a ballpoint pen, you're putting the screws to your finger flexors, powerful muscles in the forearm that become ropelike tendons, running across the wrist out to the finger bones. Whenever you want to bend your fingers, those muscles contract and shorten, pulling on the tendons like puppet strings and making the fingers curl up."

She nodded, understanding. I went on.

"Tendons pull in a straight line. Where they cross mobile joints, like in the wrist and fingers, they need pulleys to hold them down. Without pulleys, they would bowstring and lift away, robbing you of power.

"To accomplish the task, nature engineered into your hands powerful connective tissue arches to act as pulleys for the tendons, like loops on a fishing pole. The ones in the fingers are thin, belying their great strength. But the pulley at the wrist is a real brute. Securely attached and hung like a suspension bridge across the wrist, or carpal canal, it's tough enough to withstand hundreds of pounds of force. OK so far?"

Angela gulped and nodded.

"The pulleys have to be tight to work. But without lubrication, the tendons would burn up from friction. So nature surrounded the tendons in these tight spots with a delicate tissue—synovium—which makes a high-grade oil."

"Oil. I know all about oil. Is it like pasta cooking oil or body oil for lovemaking?"

I ignored the question.

"So along comes Angela, figuring she'll write eight multicarbon forms half a dozen times today. And tomorrow, and the next day, ad nauseam. For good measure, she'll throw in a couple of hours of typing each day. Typing on a computer with no carriage return. And guess what? Tendons evolved to meet the functional needs of cave people. It's only been twenty or thirty thousand years, and they haven't evolved much since then. But there was no Department of Economic Security back then—at least, I don't think so—and no computers. Life for our hands was simpler. More variety in activities, more time to rest in between. Mother Nature never figured on all these modern innovations. So the tendons get overused with no time to recover. But one mechanism, a very ancient one, still works fine—*pain.* So you hurt."

I could tell Angela was with me every step. There was one more point to complete the story.

"So your hands hurt. No surprise. The one doing the writing hurts worse. No surprise. In the beginning, time off allowed the pain to go away. But after a while, a brief time off wasn't enough to allow the tendons to recover—they need months. It didn't work anymore. No surprise.

"So you've got nine swollen, irritated tendons passing through the carpal canal. The canal, a narrow gorge at the wrist, is covered by a thick ligament-like pulley to hold the tendons down. The synovium covering the tendons is swollen and thickened, but not inflamed. Many people refer to this as tendonitis, but the term *tendonitis* is wrong, because there's no '-itis.'"

"Itis, schmitis. It hurts, let me tell you. No crap."

"But there's an additional wrinkle to contend with. In an arguably poor piece of celestial engineering, Nature stuffed into this already overcrowded space one more important structure—the median nerve. As big around as a pen, this nerve collects sensory information from the thumb, index, long, and thumb side of the ring fingers and carries it up the arm to the spinal cord, then to the brain."

Angela was a cool customer. Cocky. Sure of herself. She knew about a lot of things, but she also knew her limits. She didn't assure me that she knew about the brain.

"When the tendons swell, it's like putting veggies in a pressure cooker on the stove. The nerve sits on top of the tendons, just underneath the pulley. Swollen tendons squeeze the nerve against the pulley and cut off its blood supply. Numbness ensues. Presto. Carpal tunnel syndrome. The name for what you've got. But keep in mind there are two problems. First came tendon overuse with swelling and pain. Second, as a result of the first, came carpal tunnel syndrome with numbness, not pain. I can fix the nerve problem. But herein lies the rub, no pun intended. I can't fix the tendon problem. Time, rest, and work modification will be needed for the tendon synovium to get well. Your employer won't like it."

"My employer can take a flying leap. Won't like it, my ass. It's their fault I'm in this mess. First they work me to the bone and screw up my hands, my livelihood. Then they tell me to go see my private insurance carrier because they're not responsible. But my private insurance tells me it's work related, and punts me back to DES. So I'm stuck in the middle. No one wants to accept the responsibility for my medical problems. But you know what? I'm certain it's all caused by my work. So if we're playing Old Maid, DES gets stuck holding the card. And they damn well better make good on whatever has to be done. Right?"

"Oh, I agree. But my agreeing doesn't mean they'll give you what you want. Jobs are scarce around here and, even though they're not as competent or as experienced as you"—she blushed—"there are plenty of warm bodies searching for work, any work offering money. They're physically perfect specimens. It would be so easy for DES to replace you with one of them. They well might, if it weren't for laws to protect you. The whole scenario reminds me of movies depicting the old wooden warships. They used slaves, chained together, pulling oars to a beating drum, to row the damned things. If one slave dropped out he was simply yanked and replaced."

"Oh, they wouldn't, they couldn't . . ." Her voice trailed off and she mumbled incoherently for a few moments. Then, the synapses clicked and, from the illuminated expression spreading over her features, realization clearly dawned. Her brow knitted in an expression both quizzical and unhappy. "Would they?"

"They have a staff of lawyers on retainer. If there's a battle, you may need one, too."

I didn't relish the thought of haggling with DES. Their insurance company would probably have to approve and pay for surgery to decompress Angela's median nerves if, as I suspected, the electrical tests were severely abnormal. They really would have no choice. But even after staged decompressions and five months of treatment, Angela's tendons might not be well. If she went back to the same job, pain, not numbness, would stop her. She'd have to quit or be fired. The work was the work and I knew from past experience that no modifications were acceptable to the agency. Retraining would be needed, with additional expenses. It wouldn't go over very big.

"So I've got carpal tunnel syndrome," she said. "It's possible I may not be able to go back to my regular work when we're all through. So, like it or not, I may have to get retrained to do something else. They damn well better pay for it. Where do we go from here?" She paced the floor, walking around in little circles and waving her arms to emphasize the question. It was how she got it straight in her head, how she thought things through.

"The next step is to confirm the diagnosis. The nerve conduction test tells us the severity of compression. The EMG, or electromyogram, tells us if the motor nerve to your thumb muscles is compressed. We need to know because, if it is, your thumb muscles could become paralyzed and might not recover. A positive EMG is an absolute indication for surgery. Decompressing the nerve by cutting the pulley is the nerve's only chance."

Angela brooded for a moment. "Who does the tests—you?"

"No. I'll send you to a neurology colleague, an expert. He's done thousands. He's accurate and gentle."

"Gentle? Why gentle? You mean it could hurt?" She shifted to a look of concern.

"The EMG involves putting fine needles in your thumb muscles. You want someone slick doing it."

Angela's eyes and mouth opened widely and she backed away from me as if I had attacked her.

"Needles? No one said anything about needles. Maybe we'd better talk about this some more. Needles?"

"Relax, Angela. Art's a real pussycat. Anyway, you let them puncture your nose to put rings in. Those were needles."

"Oh—but that was different."

"Different? How?"

She threw me a condescending look and brushed me off with a toss of her head. "Men. You wouldn't understand."

As I dressed the morning of my meeting with the DES I contemplated what to wear. My power shirt was a must. But my list of imperatives petered out. In the back of my mind scratched a memory of a weapon so powerful that the mere mention of it was enough to bring all combatants to the negotiating table. To borrow an old William Powell movie line, it was like having a little atom bomb in your pocket. Even if you never needed it, it was reassuring just to know it was there. Wasn't it called the "peacemaker"? Or was the peacemaker a large-caliber pistol? No matter. Peacemaker sounded pretty good. Whatever it was, I knew I needed one of those. Oh, and a flak jacket. I expected lots of flak.

None of it should have been necessary, of course. Angela's medical problem was becoming distressingly common. Her description of her symptoms was classic. The electrical tests confirmed my suspicions. She had severe compressive neuropathy of both median nerves, worse on the left, with EMG showing advancing denervation of her thumb muscles. This accounted for their softness, loss of tone, and weakness when I examined her. She needed surgical decompression of her nerves quickly if we were going to preserve function of her thumb muscles. I knew, in my heart of hearts, her work was responsible.

I was scheduled to meet with Tony Edwards at ten o'clock at the Samaniego House Grill downtown. Tony and I knew each other. He started off as a claims representative for one of the smaller insurance companies writing workers' compensation policies. Over the years, he became more experienced and his services correspondingly more valuable. After a brief stint with the State Compensation Fund, he graduated to a managerial slot at DES. He earned a reputation for being tough as nails, a real hard-ass. But I knew better. His eight-year-old daughter, the eldest, cut her finger badly three years earlier, and he had asked the emergency department crew to call me in to sew her up. Which I did. It all came out well, and the experience was not a bad one for her. Tony and his wife, Carmella, were grateful. And Tony let me see a soft side he rarely exhibited for anyone outside the family. But this was business. I really didn't know what to expect.

"Hey, *goombah*," I yelled at him. "How's little Anthony? He must be pushing four by now. Is he running you ragged yet?" Tony looked relaxed in a casual plaid suit, his robust black hair still full and turning silver at the edges. I envied him the look of distinction. Mine was light brown and starting to fall out.

"Hey, Doc, good to see you again." We sat at a booth and caught up, exchanging pleasantries. The calm before the storm? "What's good here? I don't get to eat out much."

"I don't either," I said. "Medicine is a jealous mistress. If I want to spend time with my patients and educate them—which I do—and still finish and get home to Cindy at a reasonable hour, I work through lunch. My staff order in, or heat Chinese food in the office microwave. The smells drive me crazy, but I hold out until evening, then pig out because I'm so hungry. It's why I look like this." I snapped my belt, on its last loop. "I'm told their chicken fajitas are wonderful." We ordered and chatted about family until the food arrived. The fajitas were delicious and we devoured them without a word, then ordered more. Tony broke the silence.

"Well, I guess we should talk a bit about Mrs. Bonfiglio," he said with a sigh. Actually, although I knew she was the reason behind our meeting, I was hoping to postpone the discussion about her just a little longer. I didn't want our pleasant encounter to become confrontational. I dislike confrontations. My personality abhors and avoids them. I do have opinions, and express them when I can. But I prefer to be thought of as persuasive, not confrontational.

"There's not much to say, Tony. She has EMG-proven carpal tunnel syndrome and needs staged nerve decompressions. She's filed her industrial claim forms, and there's no question it's related to her job. What's the hang-up?"

Tony looked me over with deliberate care, as if he was measuring me for a casket. "I agree with the diagnosis. But come on, let's get serious. There's not one shred of evidence clearly implicating her work for DES as causative." He made a basket with his fingers and rested his chin in his hands. He looked annoyed. "Let her private insurance cover her. Could you pass the cream?"

"Tony, you disappoint me. Angela's a typical employee: devoted to the Department, anxious to please, super hardworking. She puts out a hundred and fifty percent and gets shafted for her efforts. I wouldn't blame her if she's mad. I would be, too. Anyway, her private insurance passed the ball back into your court. They want nothing to do with it. They say work caused her carpal tunnel syndrome and the industrial carrier has to pay medical costs and retraining. Angela's angry as hell to be caught in your battleground. You want to try this salsa? It's great."

"Yeah, please. Look, management is adamant about this overuse stuff. They don't think it will fly, and they're closing ranks to present a united front. They raise a good point. Why now? Why are we seeing so much of this stuff now? Haven't people been writing and typing for years? Hand me the salt, will you?"

"What's happening now is different, Tony. It's a combination of lack of rest—biological rest for the tissues—and this enormous emphasis on productivity. Human tissues were never meant to be worked this hard. All things fail. There's not a single product you can buy, be it a car, a TV set, a vacuum cleaner, anything, which won't break down and fail if pushed too hard. Why does it come as such a shock, such a great surprise, that the mechanical systems of the human arm fail if pushed beyond their endurance? To me, it's a bloody miracle people hold out as long as they do. You want more tortillas?"

"No, thanks. I'm gonna roll, not walk, out of here. Don't get me wrong. Mrs. Bonfiglio is a paisan. Her family even comes from the same region in Sicily I do. I believe her. But I don't have a lot to say about policy. Didn't Tom Clancy say, 'Bureaucracy mutes individual voices'? Even if I were to agree with you, even if I fought your battle in committee, I'm not sure I'd prevail."

"All I ask is for you to try, Tony. Angela's not the only one in a similar bind. I'm seeing a lot of this stuff. Ask your bosses—hell, ask yourself—would I voluntarily, willingly, do this work? I think not." Unbidden, the image of the slaves chained to their oars drifted into my mind. "Every time I testify in an industrial hearing, I try to educate the attorneys and the administrative law judges. I suspect things will get worse before they get better. You want coffee?"

"No. I have to scoot. Don't give up yet. The wheels of progress grind exceedingly slow, but progress beats the status quo as surely as a straight flush beats a full house. Call me when this one's resolved."

"Angela hasn't hired a lawyer yet. I hope she won't need one."

"I'll tell the director veiled threats were applied to me under duress."

"Oh, stuff it, Tony. At least I didn't poison your fajita sauce when I had a chance." Our eyes exceeded our capacity. There were leftovers. "Will you take these back to the office in a doggie bag? Maybe Margaret would like a tasty snack."

"I'll take them to Carmella and the kids. It's a bad precedent to be nice to your secretary, where I work."

"What an inspiring atmosphere. Next time I'll let you pick up the tab."

"Next time we'll do a barbecue at my place. Wife and kids want to see you again. We'll talk."

As Tony left, I sat at the table and finished my decaf. And pondered. Did I have good news or bad news for Angela? For sure, I didn't have any firm resolution. Like Angela, there were a lot of people out there hoping for a few morsels of good care, some glimmer of hope in their battle with the system. But where money was at stake, I was cynical enough to believe the bottom line took precedence over the bread line. Employers, defending the system, would be loath to set an expensive precedent by acceding to the validity of my arguments. But lawyers defending people with medical problems might prevail. While earning their own livings by doing so, they might (through confrontation) actually do the *right thing*. Anything was possible, especially out here in the Wild West.

It was difficult not to be confrontational with Angela. Because she wanted to be. For her, histrionics were an acceptable way of dealing with the world, diplomacy be damned. It didn't matter if responsibility for her hand function and future employability landed plop in my lap. It wasn't her style to suck up to me or anyone else. Of course, I didn't stand on ceremony where her behavior was concerned. She could throw a screaming fit, pull out her hair and dance naked with a cobra around her neck for all I cared. It wouldn't change our relationship or my commitment to her well-being. Others, though, like her boss, might not be so forgiving.

In the safety and sanctity of my office, she indulged in a screaming fit.

She paced the room, darting furtively back and forth like a caricature of the Looney Tunes Tasmanian devil. She didn't make incomprehensible grunting noises, but muttered words in an angry tone. Her arms gesticulated wildly. She allowed, no, *encouraged*, mannerisms to substitute for words.

I asked Angela if she had ever danced.

"My mom gave me ballet lessons when I was around five or six. I kept them up for eight years. Then we moved to a different town and I had to give up my classes and my teacher. I never really got back into it. Why do you ask?"

"Your movements reflect a measured poise," I said. "I wondered if you had formal training. Do you miss it?"

"Yeah, I do. I liked my classes. I liked my teacher and my classmates. They really made you jump through hoops—not literally, not like the Chinese—to learn the movements. It was work. A lot of what we learned was theory, but it was so interesting. Made me glad I was learning European-style ballet, and not that Siamese stuff."

"Why?"

"Because dance is hard enough. You trash your body just to learn to do the body movements, the flips and twists. But at least the hand movements are gentle, sort of like pantomime. You hold both hands over your heart to show you're in love. You point to your ring finger to show you're married. You hold both hands together behind your head to say you're sleepy. Like that. But that Asian stuff. They're crazy. They bend their fingers and arms out of joint to look more graceful. If I wanted to be the Indian rubber woman, I'd drink latex. No thanks."

The next few weeks passed quickly, and I saw Angela again.

"Those clowns ought to know better." Angela swept the air in wide circles with her hands as she paced. "*Marone*. Talk about biting the hand that feeds you. I give and give till I bleed, and what do I get? Nothing, that's what. No, not quite true. I get trouble, big-time. Have to have operations. Needles. Yuk. The money's good, but stuntmen get paid better and their job's probably safer. Management sit in their cozy offices and make policy while we do all the dirty work. And after years of putting out, when I need their help, what do I get?"

Suddenly she stopped pacing, turned, and looked sharply at me. "As a matter of fact, what *do* I get? It's been two weeks since you stuffed your face at lunch with that bird. I'm here suffering, no sleep, bad nerve problems, hands which might be permanently wrecked, and he's probably out living the high life, not a care, no concerns for me. Shouldn't we have heard something by now?"

She was right. We should have.

"You're right. We should have. Let me try to reach him." I

picked up a phone and dialed Tony's extension at DES. "Hello, Margaret? How are you? Did you get to taste the fajitas from our luncheon? . . . Good. He can afford to take an evening with Carmella and show her a good time. I knew you'd like them, overworked, underpaid, and as chronically hungry as you are." We both laughed. "I need Tony. Is he around?"

I nodded at Angela, who drummed her fingers on my desk while he was summoned to the phone. "Tony? It's me. I'm with Mrs. Bonfiglio . . . no, today's mail hasn't come yet . . . Seriously? That's great. She'll be pleased. Not that anyone looks forward to surgery, but she wants relief, and there's no other way to give it to her. Thanks for your support . . . What do you mean, you'll split my fee with me? Seriously, management bought the farm? Does it mean she'll be retrained? . . . Uh-huh. Uh-huh. No, she won't move to California just because mandatory retraining is the law there . . . Well, you'll just have to work on making it the law here, too, won't you? Everything else, from fashion to organized crime, works its way here from California. Why not something useful? . . . Uh-huh. Uh-huh. Yes, I'll schedule soon, before they change their minds. Call you if there's any glitch. Regards. Bye." I hung up and faced Angela with a stupid, sloppy grin on my face.

"Good news. You heard. The original's in the mail to you, with a copy to me. Your bosses caved in, a triumph of logic and reason."

"It must have been politics. They were afraid I would sue."

"You can be as cynical as you want to be, but there was no political pressure I'm aware of. Just good common sense. The bottom line is you get treated, all expenses paid courtesy of DES and its insurance carrier, the State Compensation Fund."

"You make it sound easy."

"Of course, there's no guarantee they won't hassle you or give you the runaround. In fact, it's a virtual certainty they will. If you're pissed now, wait until six months from now. If I aim you at an appropriate target you'll be ready to commit murder. But guess what? You won't be waking up at night anymore. And you'll be on the road to a new career."

"What kind of new career? I've put in years learning how to do this job. I hated school. Especially math. What am I gonna do now?" She began to sob. I gestured to tissues on my desk.

"Change isn't easy to deal with. Especially when it's in your face, and there's no escape. But don't think of it as a roadblock. Think of it as a detour—and detours are often unpaved."

"You mean I'll get back on track? There'll be an end to all this?"

"It'll be a bumpy road for a while until you get on smooth pavement again and back in cruise control. But you will. And you can give the finger to all these folks who, in truth, are just doing their jobs but who represent, to you, an unpleasant passage in your life."

Angela looked wizened, like a commander who's seen mayhem and is weary of the whole thing but sees resolution to the conflict. "Doc, I just want to get on with it. You said you could schedule my surgery soon. I have lots of questions. You'll provide answers?"

"Angela, you'll be sick of my handouts and explanations, but you'll understand clearly what's being done and why."

"Did you say they would pay to retrain me?"

"Most assuredly."

"I love to travel. Maybe I could become an airline stewardess." Her face took on a dreamy expression.

"I think you'd make a great stewardess. You'd have to get accustomed to jet-lag, but your hands would love it."

8.

Higher Authority

Connie McGinnis smashed the window, and kept shattering it until there were only tiny fragments left.

She knew glass broke with an edge approximating molecular thickness. The only thing sharper were the diamond knives she used, when her lab could afford them, in her ultra microtome. So, even in this violent outburst, she was careful not to cut herself.

"God, it felt good to pulverize something, anything. I'm so fed up with this disease. I like smashing things so much, I might go into demolition work," Connie said.

Not that it was easy for her. She had to find a pipe segment large enough for her deformed fingers to encircle and grip. It had to be long enough for safety, but not too heavy for her frail arms to lift and wield. She had to find the right tool, keep possession of it a secret, select the perfect opportunity to do the deed. It was a real production, worthy of a murder mystery.

"Where did you find the window?" I asked.

"Our next-door neighbor was tearing down his garage and rebuilding a new one. So I asked him—discreetly, of course—if I could break the window before the wrecking crew did. He knew my situation and said yes immediately. He's been a dear."

"What about your sister, Mona? Doesn't she still live with you?"

"Mona's a mother hen. She makes everything her business. She'd scold me mercilessly—if she knew. But she doesn't." Connie snapped make-believe suspenders in a gesture of pride and defiance.

"I kept my mouth shut, never told her what I planned to do, or told her I was coming to see you again. She went to the hospital to visit our nephew Jack, and pray over him. Jack broke his ankle skydiving. He's lucky to be alive."

I knew about the accident. One of my orthopedic colleagues who operated on him shared the strange story in the doctors' lounge. Jack, a second-year medical student, was dating a physically fit young woman. An experienced skydiver, she goaded him into joining her on a routine jump. Routine for *her*. The wind was gusting ferociously enough to cancel the drop if good judgment had been exercised. She landed fine, but Jack hit the hard desert floor with enough force to shatter his fibula, necessitating open surgical reduction. Good fortune must have been with him, because he narrowly missed landing on a huge century plant. If he had hit those razor-sharp radiating spines at ten miles per hour, he might have been morgue meat.

"I thought you shared everything with her."

"Not everything. You know, Mona is a Christian Scientist. So committed to it, you'd think she was Mary Baker Eddy reincarnated as a redhead. She's against all surgical procedures and believes only in divine intervention, the power of prayer, spirituality, that sort of thing. She was really upset when Jack opted for surgery to set his broken ankle, and she's beside herself I'm seeing you about my hand deformities. So, in addition to my window escapade, I have a few secrets she's not privy to."

"You were taking a big chance," I said. "The impact stress when you smashed the window might have been enough to break one of your tendons. All your extensors are badly diseased, especially the long thumb extensor. It's a miracle it hasn't already snapped spontaneously. You remember I told you about its unique anatomy. The tight tunnel it passes through, on the back of your wrist, makes it a setup for attrition rupture. It's the tendon that breaks most frequently in rheumatoid patients."

"I know, I know," she said with a petulant scowl. "My feet are bad, so I have to screw around with special shoes. My knees prevent me from walking long distances. My shoulders keep me out of the kitchen. I have more limits than the one-way streets in downtown Philadelphia. I'm sick of it. Why can't the miracles of modern medicine cure this damned disease?" Tears accented her anger, her utter frustration. Luckily, she had a stable job as lab director. She had a technician to prepare the specimens, and she was world-class when it came to electron microscopy.

I winked at her. "So how did it feel to break a window?"

She threw me a sheepish grin. "Great. Truly great. I've been wanting to smash something—anything—for a long time. This was a no-lose chance to let it out. I feel better. But my medical problems are the same." She longingly fingered the cigarette pack in her purse, but couldn't light up in my office.

"Look," I said, "nothing has changed. You remember the bit about taking inventory?"

"I think so. First comes what I've got in my hands that still functions." She gave a sardonic laugh. "With me, it won't take long."

"Right. Anatomic inventory first."

"Then, I guess, comes what I can do with whatever is working."

"Exactly. The functional inventory. And last?"

"Last comes the wish list. The things I wish I could do." She was more wistful. "The things in my life I'd give a whole lot to accomplish, if there was only a way. I want to crochet. I'd like to play the piano again. I want to make bread, from scratch like I used to, and knead the dough with my fingers." Connie became quiet, brooding.

"Ahh. The needs inventory. The hardest one, because some action is required. When you reach an impasse—something in your everyday life you want to do but can't because of your deformities—only then do we talk about surgery. I've told you many times, I don't have a cure. Surgery isn't a cure. But if you analyze the problems correctly, take careful inventory, do everything right technically—and luck is with you—surgery can help. It can. But it hurts, takes time away from your life, and it's no fun. Once you know all the details, only you can decide if it's worth it."

"Yeah. I got it. But don't rush me. This isn't easy. Mona is dead set against me doing anything. She prayed for divine intervention when her husband, Phil, began having headaches, and he went along. Still, my knuckle joints—excuse me, MP joints—are all dislocated, and the finger deviation—ulnar drift, right?—makes my hands look horrible and function worse. But I can still do my job, with difficulty. And the deformities haven't changed in a while. Better the devil you know, right?"

"Connie, it would be fine if you could count on things not changing. But you can't. One of the problems with rheumatoid disease is that the rules change in the middle of the game. It's like you're playing baseball, standing on the mound pitching, and suddenly three big guys in helmets and football jerseys rush you and tackle you. You might say 'that's not fair,' but what is? As a surgeon, I might do a fancy reconstruction and everything looks great. Then the disease becomes active somewhere else in the hand. New deformities undermine my reconstruction, and it falls apart. So you learn to have a backup plan, a fail-safe, a 'what if?' strategy. Just like in war, the secret of success is game planning. And, make no mistake, this is war." Why was I so agitated? "Look, Connie, this is old territory. We've been over it and over it."

"I know. I know you don't have a cure. I know if there was one, you'd give it to me. I know the decision about surgery is mine to make. But, damn it, I've lived with this a long time. I know what's wrong with me."

Oh, really? I said to myself as I scanned her ashen face. Connie developed rheumatoid in her early twenties and was one of the first patients to try gold therapy. Far from glittering, the heavy metal was deposited in her skin and was responsible for its odd but characteristic silvery hue. She reminded me of the Tin Man in *The Wizard of Oz*. Luckily, she'd cultivated a temperament as unflappable as Jack Haley's. Even so, if she truly grasped the mayhem the implacable autoimmune process was wreaking on her body, she'd be dismayed. This was more, much more, than just arthritis. Her kidneys, her eyes, her muscles, all her soft tissues were affected. It's why the name *rheumatoid disease* was more appropriate than rheumatoid arthritis. It makes a damned mess out of everything. The new drugs were all "poisons," including Remicade and Etanercept (initial reports seemed to show it edged out Methotrexate, at twelve times the cost). Even if they stopped the relentless progression for a time, they had side effects.

If I were a rheumatologist I would feel impotent, frustrated, feeble, in the face of an overwhelming freight train of an illness. Happily, as a surgeon, I had a few tricks up my sleeve.

"Connie, one of the biggest problems surgeons have with rheumatoid reconstructions is knowing where to start. Nothing is working normally and every bone and tendon system in the hand cries out, 'Fix me first.' From everything you've told me in past visits, finger position and pinch are vital to your ability to work. The ulnar drift has gotten so severe your fingers are deviated too far away from your thumbs to make good contact. And the rheumatoid nodules in your flexor tendons have not only paved the way for finger joint deformities but interfere with finger motion. We could talk a lot about the cosmetic appearance of your hands. Most folks, especially working women, deny they're concerned about how their hands look. Baloney! The cosmetic benefit is probably as important as any of the others. Maybe more. Your fingers would be straight."

Connie blushed. As an Irish redhead with striking, clear green eyes and freckles, her silvery skin pigmentation against her red cheeks and the intense yellow nicotine stains on her deformed fingers produced a bizarre effect.

"Think it over, discuss it with your family, and come back to see me in two weeks. If you approve, I'll set it up."

Connie was punctual. Mona insisted on accompanying her. For spiritual support, or to browbeat me? I was afraid it was the latter.

"Dr. Arem, this is my sister, Mona."

Mona was a redhead like Connie, but a different shade. Razor thin, with rather severe makeup, she came across as a stern version of her younger sister.

"Hello, Ms. Spencer. Nice to meet you," I said.

"Please—Mona. I prefer first names. Doctor, let's get right to the point. You want to operate on Connie. That's your bias. You think it's worth disrupting her precarious schedule and committing her to months of uncertainty and helplessness, especially if you do both hands in sequence. I think your bias represents a distorted view of reality."

"Please go on," I said.

"Connie's in pain all the time. She's learned to live with it. I pray every day her pain will go away and, through spiritual understanding, she'll heal. If she insists on having surgery, the end result has to be worth it. Let's talk about the outcome. I know I'm abrupt. But I've learned it gets me answers. Connie, honey, am I representing you fairly?"

Connie seemed disturbed. "Mona, I've had the benefit of Dr. Arem's explanations. You haven't. I'm a competent person. I think I'm capable of making judgments like this myself."

"Nonsense," said Mona. "When Phil died and I moved in with you, you and Jack were barely managing. I've watched your hands get worse over the last three years. You wanted to run the household. Not that you ever liked cooking to begin with, but you can hardly lift a pot of water now. I'm happy to take over the kitchen. If you want to be a martyr, let them buy you out of the laboratory, with all those solvents. I get a headache just thinking about it."

"Nothing like the headaches Phil had before his aneurysm burst."

Mona fell silent and merely nodded her head.

"One other important consideration," I said to Connie. "You have both severe MP joint disease with all the trimmings, as well as terrible extensor tendon inflammation. I can only fix one problem or the other in one surgical sitting. More would take too long and be too much surgery. So which do I tackle first?"

The two ladies looked at each other and shrugged. Connie was wavering. On the one hand, she was strongly persuaded by my detailed explanations. But on the other, she had great respect for Mona. Even though a purely spiritual approach was antithetical to her personal belief system, part of her wanted to believe. And there was no doubt in my mind a positive, believing, and upbeat attitude toward her future would be good for Connie, no matter what treatment was offered her. Hope is a powerful force. The most compelling reality she could cling to now was the reality of her daily pain, deformity, and limitations. If a combination of surgery—mechanical manipulation of tissue—and a spiritual slant was beneficial to her, I was all for it.

"My colleagues and I have found out, the hard way, that it's better to do the MP joints first, rather than the other way around. It's more conducive to the extensor tendons moving postop, and the results are better. Because your thumbs are in reasonably good shape, I'll try to sneak in and remove enough synovitis from the long thumb extensor to eliminate the risk of tendon rupture."

I finished the explanations, and ended by showing Connie and Mona some before-and-after photographs of another patient with similar deformities. By this time, Connie was attentive, sitting at the edge of her seat. Mona seemed numb. I've discovered exhaustive, thorough explanations, even delivered in understandable lay language, can sometimes be too much.

The next day, Connie called to ask me to schedule her operation.

Jack became intrigued with surgery as a first-year medical student. Single and unattached, he hung out in the ER after hours and wangled his way into the operating room. At first, his presence was tolerated as a curiosity. He didn't mind being the butt of the often vulgar OR humor. He knew it would change. After a while, he was invited to scrub. Then he was asked to help. Eventually, his presence was not only tolerated but welcomed. Many times he first-assisted the surgeon, allowing the scrub nurse to do the job for which she had been hired.

His ankle fracture slowed Jack down a bit and kept him off the basketball court. But his newly acquired skills were undimmed in the operating room. Perhaps because of his Aunt Connie's predicament, perhaps because he was haunted by the specter of a familial tendency and wondered if he was at risk, hand surgery held a special fascination for him. So he scrubbed on hand cases whenever he could. For an added bonus, he didn't have to stand holding retractors, the way he did in bowel cases. Hand surgeons get to sit. As with eye surgery, delicate work demands steady hands. We would all need them. We had to carry out procedures that would tax our skills to the limit.

MP arthroplasty: the way we were going to do it, it's one of the most challenging procedures imaginable. A lot of the challenge is due to the complexity of rheumatoid disease.

Joints are named for the bones they separate. There are three phalangeal bones in each finger. The metacarpals are the large bones forming the center of the hand. The MP or metacarpo-phalangeal joints, commonly called the knuckle joints, connect the metacarpals with the phalanges.

Connie's MP joints were severely diseased. Displaying a deformity called ulnar drift—common in rheumatoids—her fingers angled toward the ulnar, or little finger side of the hands. This was striking, visible. Underlying it was complete destruction of the MP joints and a host of subtler problems, all of which had to be recognized and corrected or her reconstruction would fall apart. And we only had one tourniquet time—about two hours—to safely get it all done. Jack knew he lacked the skill and experience to first-assist, and bowed out to Susan, my regular assistant, who had done more than fifty of these with me. But I was happy to have Jack sitting in as second assistant. All of hand surgery requires precise retraction of skin and tissues in an operative field the size of a postage stamp. The retractors are small, and skilled help is invaluable.

We were using a block anesthetic, so Connie was awake but sedated. The block was perfect, so she was free of pain, a novelty for her. The surgical drapes made a dark cocoon over her head, and she snoozed.

The incision was transverse, across the knuckles from the index to the small finger. I scored the skin to mark the path, but only deepened the incision for the joint I was working on. One of a myriad technical details, it prevented the tissues not yet being operated on from unnecessarily drying out.

The sequence of steps was well rehearsed. First, split and retract the extensor tendon. Remove the inflamed lining from the MP joint and tendon undersurface. Dissect and preserve what was left of the radial collateral ligament, the ligament providing pinch stability. Use the microsaw to remove the correct amount of metacarpal, relieving the joint tightness. Fish out the flexor tendons and remove inflamed tissue enveloping them. Use a reciprocating rasp to make space in the bones for the silastic implants. And on, and on. You have to see it to appreciate it. Over the years, I've become accustomed to doing impossibly complicated operations and making unreasonably difficult judgments almost every time I'm in the operating room. Sometimes it's good to have observers to remind me of this and encourage me.

"These silicone implants you're using are just flexible hinges, aren't they?" asked Jack. "They're not really artificial joints."

"Good for you," I said. "As you saw, the whole key to the reconstruction is to remove enough bone from the metacarpal head to fully correct the dislocation, eliminate tightness, and rebalance the joint. There are a lot of technical details, all of which are important. But you're absolutely correct. The implants are spacers, keeping the bone ends apart so they don't fuse, allowing a measure of reproducibility not possible before Al Swanson developed, and Dow Corning produced, these implants. OK, we've reconstructed one digit. Tired as I am, and much as I'd like to stop now, we still have three fingers to go. Let's get on with it."

Connie had been dozing. Now she woke up.

"Connie, I have a friend of yours with me."

"Hi, Aunt Connie."

"Jack. Is that you? You do get around. I've been saying lately, what with your terrible study hours and not taking good care of yourself, you've been getting under my skin with me worrying about you. It never occurred to me you'd take me literally."

"I hope you don't mind, Aunt Connie. This is really interesting."

"Of course I don't mind. Actually, I feel relieved, knowing you're there to keep an eye on things."

"Thanks, Connie," I said. "Thanks a lot."

"Any surprises?"

"No," I said. "It's what I expected. We're finished with the index finger. The MP joint was completely destroyed. I had to pull out the flexor tendons and remove some large inflammatory nodules. You'll have better movement now. I'm not happy with the collateral ligament, though. There was almost nothing left to reconstruct. We'll have to protect you against deviating stress for eight weeks. I know it's corny, but I can't help reiterating the bad joke: "What's a joint like that doing in a nice lady like you?' I've got a big implant in, and size should help with stability." Connie should have hissed at my bad joke, but she said nothing and focused instead on Jack.

"Jack, you came in way after hours last night. We have to have a serious talk, young man. Your behavior is an abomination. It's unacceptable."

"Aw, Aunt Connie. I'm a big boy. There's nothing to discuss."

"In a pig's ear, there isn't. We established rules when you came here from Boston to live with me while you went to school. We had an agreement. It's not fair to me. I'm not sure . . . I'm . . . I'm . . . mmm."

Alarmed, I glanced at the anesthesiologist. She nodded knowingly and gave me a thumbs-up sign, so I would know she had taken it upon herself to slip Connie just a tiny bit more sedative. I didn't argue with the imposed stillness. I had quite a bit more work to do, and needed to concentrate.

Jack, looking sheepish, said nothing.

Connie came to my office for her first dressing change.

"Well?" I said. "Talk to me."

"Piece of cake," Connie said, smiling.

I was glad she thought so. In twenty years of practice, I've never used those words to describe a procedure to a patient. Even when I felt they were accurate. One never knows.

"Mona's not talking to me, but she'll get over it. She always does. Everything else is going great."

"Pain manageable with the medicines I gave you?"

"Not bad at all. The pills helped the arthritic pain in the other hand, too. But look—I can wiggle them fine. And see—they're straight."

Connie beamed. The expression of glee in her eyes told me she was thrilled with the cosmetic improvement, even before she got to actually use her hand. The hand therapists would put her in a protective bandage, with a carefully made splint and outriggers to control her motion and finger position.

"When do we get the other hand done?" Connie said.

"When you can be comfortably one-handed with this operated hand."

Connie came in with Jack four months after the second side was done. Jack had his final cast off and was getting around fairly comfortably. No limp, I noticed.

"Well," I said, "how do you like your hands now?"

"Just as you told me they would," she said, "they certainly look better. They're still pretty weak. But I can actually pinch things with my thumb against my index finger. I couldn't before. And I've done some crocheting. It's funny how you get used to limitations. At first they seem monumental, gargantuan. Then you adapt, use what you have and try to live a normal life, if you can. When Phil died, Mona was like a crazy woman for a while. She couldn't get used to the idea of being alone. Not that Jack and I can substitute for him, fill that void. But she made a conscious trade: her help with chores for our company. Say what you want, it's worked okay. Jack's antics drive us both bonkers, but he'll grow up someday. Won't you, Jack?"

Jack aimed his cherubic face at us and smiled, blinking his eyes and long lashes innocently.

"And what secret have you kept from Mona?" I asked. "You mentioned it but didn't elucidate."

Connie was silent for a time. When she spoke, it was with a conspiratorial whisper.

"I'm writing a memoir. It includes some juicy anecdotes about Mona, our life together, some reflections I've had. I don't have any illusions about chances for publication. But I think the information might help someone, and writing it is cathartic for me."

Jack apparently sensed an opening and made his move. "Aunt

Connie, could I use the house for a class party? You'll get to meet a lot of my friends, and Dr. Arem's invited, too. They're pretty neat, and some of them will stay in town to practice when they finish residency. You'll have a ground-floor relationship with your future doctors." Jack's expression was hopeful, pleading.

"It has to pass muster with Mona," Connie said. "She's the one who'll have to make refreshments and clean up the mess afterward. Although I'll be happy to bake some bread for the occasion. Have you broached the subject with her?"

"No." Jack groaned. "I was hoping to avoid it. Can't you exercise executive privilege or something? Claim the decision comes from a higher authority?"

Connie blushed again. "Jack, I can make a lot of claims. Power to influence higher authority isn't one of them. Just look at me. If I had any pull upstairs, you don't think I'd choose this, do you?"

"I love you, Aunt Connie. I'll go pester Aunt Mona about this party."

Connie looked at me with an expression of perfect bliss.

9.

Perfection

"Don't eat that pencil, Ricky. It's yucky. Check it out." Alberto's tone was stern, but his eyes were filled with love for his baby brother.

Ricky smiled and adjusted his position on his chair. He blinked a few times, his long eyelashes accenting deep brown eyes. An appealing two-year-old, a bit roly-poly. His mother had the same tendency. Although she wasn't here today, she had come in to sign some forms a few times, so I had had a chance to observe her. Considering the magnitude of her drug use it was no surprise her weight stayed down. I suspected she might have been seriously overweight but for her cocaine addiction.

Ricky reached for the pencil with his right hand. Good, I thought. More confirmation of what I observed when I surgically separated the fused fingers of his left hand last year. He was born a righty. Hand dominance would be a bitch to change, and I was happy we wouldn't have to. As it was, Square and Compass, the state-supported funding agency, would be picking up the relatively huge tab for the rest of his reconstructions. Adding a therapy bill for the marginally effective work of changing hand dominance would probably strain budgets to the limit. Ricky would need hand therapy, a lot of it, for many years as part of his rehabilitation. I could only hope his predicament, plus interest in working on his unique deformities, would stimulate the therapists' generosity.

Nor was making a thumb part of the daily routine for the

average hand surgeon, either. Except for those few individuals who pioneered a technique and needed patients to prove its worth. The way to achieve immortality in medicine is to have something named after you. So there was the "Riordan" method. Or the "Littler" method. The lack of thumbs at birth is an uncommon deformity. It's becoming even less common with growing awareness and avoidance of the numerous causes of birth defects, such as taking thalidomide and many other drugs in the first trimester.

But knowing about the harm drugs caused didn't stop Estella, Ricky's mother, from shooting up and snorting lines of coke. Ricky's birth defects spoiled the fun for her. Ricky needed medical care. He couldn't simply be swept under the carpet and ignored, like Alberto sixteen years earlier. Now Estella was married and could not so easily behave like an irresponsible single parent.

"Ricky's being agreeable today," I said. "Is he always like this or is there a dark side?"

"Mostly, he's a pretty good little guy," said Alberto, nodding, running a hand through his thick brown hair. Alberto's face was long with high cheekbones. "He has his moments, like the rest of us. He screams real loud when he gets pissed off. Aunt Rosa's pretty tough with him, though. She don't take no crap, you know? I mean, he's got reason to be frustrated, tryin' to pick up things with messed up hands. But he does OK with his fingers. I guess he's used to it by now."

I watched Ricky hold the pencil. Lacking thumbs, all he could do was squeeze it between his fingers, the way you would hold a cigarette. "Cigarette pinch" works in the absence of an opposable thumb. Ask any monkey, he'll tell you. It's great for swinging from trees, especially if your arms are fantastically strong and long enough to brush the ground when you walk upright. But try, without a thumb, to write, to use tools, or to play a musical instrument. If you're a human being you do the best you can with what you have. But it's limiting. *Disabling* doesn't begin to describe life without thumbs; frustration is just the beginning. If Ricky could speak he would weave an endless tale of difficulty. My goal was to make thumbs for him at a young age, before habits of hand usage became deeply entrenched. No small task, because we begin using our hands to manipulate our environment from infancy. And Ricky had other hand deformities to attend to.

"Rosa's pretty important in your life, isn't she?" I asked. "I hear you talk about her much more than your mom."

Alberto's expression was chilling, vicious, hostile. "My mom's been out of the picture for as long as I can remember. I don't know how or why my stepdad puts up with it. I mean, what's in it for him, you know, man? He supports her habit, and it's pretty *expensivo*. Maybe she gets money from some state or federal program I don't know about. I don't wanna know. I just want her to stay away from me and Ricky. Far away. She's bad news. If it wasn't for Aunt Rosa I'd be in deep shit. And Ricky wouldn't have a prayer of gettin' his hands fixed."

Ricky ignored us. He was intent on making a design with the pencil. His face was a mask of grim concentration.

Alberto smiled, appearing content but thoughtful. "Mom finked out on me, but she's not gonna mess up Ricky's chances to have a life. I been in church with Rosa, I felt the power of her prayer. Everybody deserves a chance. Ricky, just like me. Only he needs a lot more help than I did. Special help. Rosa helped me, brought me up. Without her, I'd be nowheresville. If I can, I'm gonna help Ricky. I need you to tell me what to do, Doc. I'm only sixteen, but I got a job hustlin' burgers, an' I got wheels. Aunt Rosa said she'd sign stuff, since Ricky is illegitimate. It'll be interestin' to see if my stepdad, Benito, sticks around. He's been supportin' Estella for four years, but now he's makin' noise like he wants to split. Neither one of 'em can be counted on where Ricky is concerned."

Ricky heard his name mentioned and grinned. But he remained intensely focused on the squiggles he was creating. He babbled happily to himself.

"I admire your dedication to Ricky," I said. "He's going to need me to make him a right thumb soon. Like before, when I separated the fused fingers on his left hand, I'll need authorization from a parent. Even though you're assuming responsibility for his welfare and upbringing, legally I still need your mom's signature. Do you foresee a problem?"

"Naaah," said Alberto. "I'll sic Rosa on her. Rosa's her real aunt, and madder'n hell about the drugs, the carryin' on, an' not keepin' up her end. Especially where her kids are concerned. In our culture, in our family, you just don't do that kind of thing. Rosa'll put the fear of God in Estella. She'll sign. *No problemo.*"

"How are you holding up, Alberto? You're carrying a heavy burden."

Alberto's face wore an expression making him look older than his years. But his expressive brown eyes were intense and bright. "I'm OK. Rosa an' me, we got an understandin'. The Bible has answers to all our questions. If we let God into our lives, everything comes out right. When Mom dropped me, Rosa picked me up. I been blessed ever since. You know?"

Knowing Estella, I couldn't disagree. Alberto was a tough kid, strong with good values thanks to his great-aunt. Maybe it was a blessing, Estella giving up on Alberto's upbringing. Who was to say? Alberto was well fed, well clothed, and had high moral standards thanks to Rosa. A decent human being, committed to helping his brother. If Estella had held on to him? Knowing her, I doubted Alberto would have fared as well, in any category. I vowed to give Alberto all the help I could.

"We're coming to a major crossroads in Ricky's life," I said. "He's right-handed and I need to make him a right thumb. Luckily, the fingers of his right hand are all normal, not fused together as they were in the left hand. I'm going to use the index finger, rotate it and shorten it to make a thumb." I fidgeted nervously. "It's four hours of pretty complicated surgery. I need written authorization from your mom—she has to sign the consent. Can you arrange to get her in here? It would probably be good if Rosa came, too. Any problems?"

"Naaah. It'll be a pleasure. Leave it to me. Rosa wouldn't miss this for the world. Mom likes doctors' offices. It wouldn't surprise me if she tried to hit you up for some narcotics."

"Yeah. Right. Except I don't have any." My God, I thought. What was I dealing with? At that moment, I felt grateful for Alberto's strength, his willingness to intercede on Ricky's behalf, for acting as a buffer between me and Estella. Because I wasn't at all sure I could keep my cool if I had to deal with Estella alone. It would be all I could do to be civil to her.

Ricky remained engrossed in his squiggles.

It took more than a month to solidify the arrangements. Even though the others were pliable and cooperative, Estella wouldn't commit to a time. For Alberto's sake, I wanted to avoid confrontation. He had, consciously and willingly, assumed the role of surrogate; to be for Ricky all the things Estella was not. Was she grateful, or resentful? For that matter, did she have any feelings about it at all? Did she care? My anger, and I had plenty, was beside the point. My assumptions about the role and responsibilities of motherhood vanished to insignificance next to the reality of Alberto's sacrifice. If it were his decision to make, Ricky would have been in the operating room long ago. The injustice lay in the blind dictum of the law, which recognized Estella as mother.

I was going to have to provide reassurance, to take center stage and be exemplary. Calm. Professional.

Added to my misgivings, to my nervousness, was the simple fact that I had never before made a thumb out of an index finger. I knew how, to be sure. I had even obtained, from the American Society for Surgery of the Hand, a teaching videotape made by German hand surgeon Dieter Buck-Gramcko on his technique for these reconstructions. So I knew how to proceed. But I had never done it before.

I asked my colleague Dr. Robert Evans to be my assistant. A fine hand surgeon, he had never done one, either. But he was competent, good help, and very supportive. He had also reviewed the tape. We were well equipped to get the procedure done with finesse and do Ricky justice.

The stakes for Ricky were enormous. He had only one dominant right hand to use, for everything, for the rest of his life. If I screwed up, that was it. Neither of us would get another chance. I was going to use his index finger, shorten it to the length of a thumb, and rotate it a full ninety degrees. I had to take tendons used by the index for one function and revamp and reconstruct them to do entirely another. I had to make index finger joints work differently, like thumb joints. I had to create functional skin flaps seemingly out of thin air, packaging the whole reconstruction to not only appear right but have a thumb web space that was aesthetically correct, folding up out of the way when he opposed his new thumb against the other fingers but ample enough to pull his new thumb widely away from the fingers to grasp large objects. And I had to preserve normal feeling to all areas of skin by preserving and protecting the tiny nerves. And make the whole thing stable. And make it look good.

It was a bizarre assemblage in my office that day. Estella, true to her reputation, looked like a prostitute in a short, tight black skirt, black heels, a provocatively clingy sheer blouse and gobs of heavy makeup. Her husband, Benito, was pleasant in a tan suit. He quickly retired to the background, and seemed ineffectual. He fawned over Estella and contributed nothing.

I had never met Rosa, only heard about her from Alberto. Lean and thin, she hovered in a worn-out floral dress. Her burning green eyes absorbed everything, missing nothing. I trusted we were on the same side, because I would not have wanted to cross this lady. She exuded a calm certainty of purpose that reminded me of Charlton Heston's Moses sizing up Queen Nefertiti.

It was a chess game, and Estella made the first move. A pawn's move.

"Let's get this show on the road. I got things to do. Important things."

"What could be more important than Ricky's welfare?" Rosa took up the gauntlet, her voice icy and brittle. Estella glowered at her.

"I didn't say more. Just as. At least, to me."

"As you say, dear, let's get the show on the road. Why don't you go ahead and sign the consent for Ricky's surgery, and we can all go about our business."

"Not so friggin' fast. How do I know it's goin' to be done okay? An' who's goin' to take care of him after? Won't there be lotsa visits to therapy, and all?" She looked at me with a sneer. "That stuff is expensive. An' the surgery's so complicated, I can't even picture how you're goin' to pull it off." She finished her tirade, seemed pleased with herself, and focused intently on her fingernails.

Alberto understood his mom's hidden agenda. She was worried there might be unpaid bills and she would get stuck with the tab.

"The doc's told me all about it. He knows his stuff. Everything's been cleared with the funding agency and with therapy. It's all covered, a hundred percent. Rosa an' me'll take care of Ricky. I'll see he gets to therapy when he has to."

"Oh, so my little man's goin' to be a mommy to the gimp. How sweet. Where's the payoff for you?" Estella, it seemed, thought some money would be provided for child care and she coveted it for herself, mistakenly envying Alberto for his windfall.

But Alberto was spared the need for a typically modest and humane response. Rosa intervened with a stern tone: "God will see to Ricky's improvement. You're a disgrace, Estella. You're Ricky's mother. At least, you gave birth to him. All these years you haven't lifted a finger to help him, except to sign on the dotted line so others could care for his needs. You bring shame to this family."

"And you're a hypocrite, Rosa. Don't give me that Goody Two-shoes line, tryin' to make me feel guilty or somethin'. I can see right through ya. Who's kept you so cozy and comfortable all these years? You been suckin' up to that nice pastor? Is he the one been payin' your bills? Or is it your fee? Is it just him, or is there more 'n one?" Estella's leer was accompanied by obscene gestures. Alberto's jaw worked silently, clenching. Rosa preempted his response.

"Estella, you make me sick. I feel shame to count you as part of this family. Your mother, my beloved sister, may she rest in peace, would also feel deeply ashamed if she were here. I—"

"Oh, cut it, Rosa," snarled Estella, jumping to her feet. "You don't have any exclusive on saintliness around here. I'm the one who has to sign. I'm the one with the power, so don't you forget it. I could just snap my fingers an' the rest of you do-gooders would have to shape up an' jump to it."

While Estella gestured, Alberto stood up and turned to face her. As he fixated her with a withering stare, Estella froze. He pointed a menacing finger at her.

"OK, I've had just about enough." He breathed heavily and spoke through clenched teeth. "You done as much damage as we're gonna put up with. You messed up my life, but Rosa taught me forgiveness. It's not okay—it'll never be okay—but I'll let it go." Alberto exhaled forcibly and fought for control. "You smeared Rosa's good name. There's no excuse for that, especially comin' from scum like you. She can forgive you if she wants. But you ain't gonna mess up Ricky. He's got a shot, an' you're gonna do the right thing now, if it cripples you. If it don't, I'm goin' to . . ." As Alberto took a threatening step toward her, Benito suddenly materialized, blocking his path. Before he could advance further, Rosa grabbed Alberto's muscular arms and eased him backward.

"Not for us to judge," she said. "Judgment, when it comes, will be swift and sure."

"Sign," hissed Alberto. "Before I lose it."

Estella looked around at the angry faces and, with a backward toss of her head, laughed and quickly signed the consent forms. "Come, Benito. Let's get out of here. I need some fresh air." She grabbed his jacket sleeve and pulled him out of the exam room. I heard the outer door close with a resounding thud.

We all looked at each other. The atmosphere was stifling, still murky. Rosa was the first to smile.

"Well, I don't know that we resolved anything. She'll come to a bad end, that girl. But she signed. Doctor, go do your thing. If we all do our part, and do it with love and clarity of purpose, how can anything but good come from it?"

I had no answer. I was still shaking from emotion. To myself, I said a small prayer of thanks that they were too busy feuding to pin me down about my experience with the reconstruction. I looked at Alberto.

"We have a deal?"

Alberto nodded thoughtfully. "We have a deal."

"Let's do it before she changes her mind. I don't trust her any more than you do. At least she never asked me for narcotics."

Alberto grinned a lopsided grin. "We never gave her a chance. But don't count her out yet. Bet you a twenty she'll think about it and come back."

"With luck, I'll have Ricky's surgery done by then. You were pretty formidable in there. It took guts to go against your own mom."

"I have a mom, and she ain't it. To me, she's just a not-so-nice person who pissed me off. Life's too short to have to deal with people like her. I got too much to do."

The day for surgery arrived with alarming speed. At least, alarming to me. I was as prepared as I was ever going to be. Though the procrastinator in me still wanted to put things off, if only for a little while longer, I sighed with resignation, put on a mask of confidence, and showed up early to organize and orient my OR crew. Never having seen it before, they gawked in fascination at the deformity, mercifully directing their attention away from me.

Evans met me by the scrub sink. "I'm impressed. Pretty normal hand, except for the absent thumb."

"Yeah. The left hand's going to be a bitch when we get to it because of the scarring from the previous reconstruction. At least, on this side, fashioning the skin flaps right should be easier. It's the key to appearance of the new thumb. You sleep well?"

"Probably better than you. I'm just helping. You carry the responsibility." He was right. He knew from experience. Our roles were reversed often enough.

"I'll do the flap and nerve dissection first, expose all the muscle units and get them ready for transfer, and remove the metacarpal to achieve our shortening," I said. "That'll take two hours and we'll probably take our tourniquet break then. It's a small hand. I'm going to wear my magnifying loupes. You?"

"No," he said. "My eyes are pretty good at close distance and I can be better help if I have a wider field of view. I'm just holding retractors. You're the one with the scalpel, remember."

"Yeah. How could I forget?" It was too late to run.

Bleeding is the enemy of the hand surgeon. Of course, if you get cut and don't bleed, the majority of the time it means you're dead and, therefore, not a good candidate for surgery. I mean uncontrolled bleeding, bleeding that gets ahead of the surgeon and stains the tissues red so it's hard to see landmarks.

Hand surgery is done using a pneumatic tourniquet, a special device like a blood pressure cuff that squeezes off blood flow to the arm at a controlled pressure. When work is under way and blood vessels are cut, there's a little bit of bleeding from blood still left in the arm. We control this with a bipolar cautery, needle-sharp tweezers connected to an electric power supply so minimal current flows only between the tips of the tweezers and doesn't spread into the tissues.

Contrast this with a conventional cautery, a small piece of metal shielded where it's grasped, attached to an electric power box. Typically, the surgeon clamps a blood vessel and touches the metal cautery tip to the clamp. Current flows from the cautery down the metal clamp, out into the tissues toward a grounding plate, usually beneath the patient's thigh, completing the circuit. The electric current fries everything in its path, concentrating in areas where metal touches tissue. Tissues held by the clamp boil and smoke, smelling like burning chicken feathers. At a cellular level, this is like strafing with napalm. I imagine the scene in the movie *Independence Day*, in which the alien death ray moves sequentially down the floors of the Empire State Building, blowing out the walls and windows amid flames and screams. I shudder at the thought.

A bipolar cautery is nondestructive. It's not only possible but routine to coagulate bleeding vessels on the surface of a nerve without harming the nerve. Try it with a conventional cautery and you cook the nerve to a crisp. Not conducive to function or good results.

Tiny patients mean tiny arms. Tiny arms mean tiny blood vessels, so I was happy to be wearing my high-power magnifying loupes. The loupes also made it easier to see and sort out all the structures I was rearranging. Which meant just about everything. Bless the Buck-Gramcko videotape. It provided context, allowing adjustment to orientation and anatomic detail. I've always wondered how orthopedic surgeons could move sequentially from doing a hip replacement to repairing a nerve. There's such an enormous shift required in size, scale, muscle force, and instruments used.

I finished the first phase, which I term the destructive part, precisely on time. After a safe limit of two hours, metabolic toxins build up in an arm deprived of blood circulation, and it's necessary to apply a temporary bandage, take a break, release the tourniquet, and let circulation resume for twenty or thirty minutes. It's then safe to milk the blood from the arm by wrapping a tight rubber bandage from the fingertips to the tourniquet cuff, reinflate the tourniquet, and go another two hours. I've actually done this three times in a complicated tendon-nerve operation.

I released the tourniquet and was gratified to see all the fingertips immediately turn pink and healthy. The bipolar cautery had done its job, and there was almost no bleeding. The hand had changed dramatically. Even without me completing the reconstruction, what was once the index finger was now shortened and rotated, looking very "thumblike." I sighed deeply, stretched, and went to the doctors' lounge with Bob Evans for decaf coffee (no caffeine shakes, thank you).

"Well, Bob, what do you think?"

"I like it. You make it look easy. I'd never guess this was your first one."

"That was sure the right thing to say. Whether you're sincere is something we'll talk about over a beer later."

Actually, though, it was going well. Now that we'd taken the jigsaw puzzle apart and scrambled all the pieces, we would get to lovingly put it back together. The new way. I hoped Ricky would like it.

I hoped he'd learn how to use it. More than anything, I think, I wanted him to be able to hold a pen and write.

I told Bob my fond hopes.

"Will you still like it if he writes pure drivel?" he said.

"In a word—*yes,*" I replied. All I could do was give him the tools. What he did with them was up to him.

"Looks to me like you've got it made. Are the jitters subsiding?"

"Bob, you know as well as I do there's always something to fret about. Even if there wasn't one, I'd invent a problem."

"Perfection is the enemy of good. When are you going to learn it?"

"When you do."

"Zingo. You've got me, I guess. Finish your coffee and we'll go button it up. Should I ask Cathy to come in and give you a back rub? It wasn't part of her nurse's training, but she's a pro at it, and she's working in clean corridor today."

"No. I'll get too relaxed," I said. "I need some tension to keep me alert. I'll ask Cindy to do it when I get home tonight. Not that I'll need help sleeping after this is all finished."

"First-time jitters humble us all. By the time you've done at least one of everything, you'll be too old to operate safely. And that's not even considering the new operations you'll invent to solve problems as you go."

"Zingo. My turn. Shall we?"

Bob gave me a reflective stare. "You lead. I'll follow."

The end was in sight. I didn't need a second invitation.

Back in the operating room, with the tourniquet reinflated, I took careful inventory. Was I forgetting anything? No—I had jotted down the important details on a draping sheet with a surgical marking pen before we took a break. So I was ready. The final ninety minutes were a blur of adjusting, suturing, endlessly modifying. I felt like a garage mechanic fussing to get the timing perfect. Then Bob's admonition hit home and I realized it was time to stop. As I tailored the skin flaps into position, the tension sitting on my shoulders like a great beast finally got off, leaving me fatigued and strangely buoyant. Bob sensed the change.

"Well, this one's a winner. What will you do for an encore?"

"There's always one more mountain to climb, isn't there? Trouble is, you can't see the next peak until you get to the top of the one just before." And the great crush of bodies in need never ends. Open the floodgates and you get trampled by the frenzied rush of people hoping to use your expertise before you disappear. I had the experience when I volunteered my services at the St. Elizabeth's free clinic.

"So what's the answer?"

"For me, it's always been to take care of people one at a time. And not let urgency, or monetary pressures, or bureaucracy get in my way." When I go into my private exam room with a patient, it's as if I entered a time warp. I shut out the world. All irrelevancies are suspended. It's just me and the patient and the problem, nothing else. For as long as it takes. I like it that way. I'll fight with my last ounce of energy to preserve it. In Ricky's case, we still had the other hand to do.

"You'll let me help with the left thumb reconstruction?"

"You're my man. I wouldn't have anyone else."

Bob seemed content. I could go out now and reassure the family that all was well. The bulky bandage encasing Ricky's hand wouldn't let them see much, but the tips of the fingers were exposed to allow circulation checks. Instead of four fingers lined up in a row, there were three with the fourth stuck way over opposite them where a thumb should normally be. It wouldn't take mental wizardry to figure out a major transformation had occurred, and they would see the final result soon enough. But I realized I would have to interact with Estella, the wild card in this mix. It was irrational, I know, but anticipation of any meeting with her unsettled me.

Three months later, we all met in my office again. Ricky was playing with a toy whose moving parts required breakdown and assembly. Damned if he wasn't using his new thumb to do it. What would be trivial, not even worth mentioning, for a child with normal hands was, for Ricky, a triumph. Yet the ability to manipulate the toy was almost assumed. Just as breathing difficulties during a cold disappear effortlessly when the virus has run its course, use of a thumb for a child is built into the program. If it's there, it will be used. For Ricky, it was as if not having a right thumb at birth was an impertinent and temporary glitch.

Estella was clothed more conservatively than at our last meeting. A plain green dress. Tan flats, no nylons. Subdued makeup. Was it for my benefit? I wondered. Was she angling for something?

Alberto seemed elated. He played with Ricky, offering him rattles and games, smiling broadly each time his baby brother grasped and held on to the toy. Neither boy tired of the experience.

Rosa, cool and detached, watched, Sphinx-like, from a corner of the room. Did she, also, mistrust Estella's motives?

We didn't have to wait long to find out. Estella began to wince, furrow her eyebrows, and rub her forehead with a look of worried concern. She turned to Benito, who massaged her temples in a soothing pattern.

"Doctor," she said, "you did a wonderful job. Ricky has a thumb, thanks to you. We're all so pleased. When do you wan' to work on the left hand?"

"There's no reason to wait," I admitted. "Soon. In about three months. When he can be one-handed with the right hand and the reconstruction is strong enough to handle anything he throws at it."

"Fine," she said. "I'll sign the forms now, if you want. I'm sure it'll work out well. No reason to wait."

Uh-oh, I thought. Here it comes. First the carrot, now the stick.

"By the way," she said in a distracted tone, "I been havin' these terrible headaches. Must be the excitement, or strain or somethin'. Aspirin an' Tylenol don't touch 'em. Do you have anythin' stronger? I jus' need a few." She eyed my sample closet, its door closed, as if it were a heavily guarded vault housing a fortune in gold. Alberto, within earshot, wore a knowing grin.

I snapped my fingers, shaking my head slowly. "Gee, Estella. I had a few Lortab samples, but they outdated and I threw them out— last week." I made it far enough in the past to allow for a trash pickup. I didn't want her rummaging through my garbage bin outside. "The cupboard is bare. I'm sorry, I know a good acupuncturist, though. . . ."

"Ah, no, that's OK. I gotta be goin'. Come, Benito." She motioned urgently toward the door.

"I'll have my secretary assist you with the forms. I'll notify you before the next surgery." I walked her out and came right back. Alberto, tight-lipped, nodded.

"I should've bet you that twenty when I had the chance. A little pocket money'd come in handy."

"No chance," I said. "I expected her to try something. I'm surprised and pleased she didn't make more of a scene."

From the corner of the room, Rosa's voice intoned softly, "She could no longer figure out a way to benefit from Ricky's deformities. So, being a selfish person, she gave it up and moved on. Like a cyclone that weaves a path of destruction and suddenly changes direction. Let us all gain strength from the calm which follows the storm. Doctor, Ricky will have a chance at a normal life. Love surrounds him. He is blessed."

Rosa was blessed, to be able to see things in black and white. To me, at least, the picture was a bit muddier than she painted it. But what the hell, we all have limitations. We do the best we can with them. In a dysfunctional world, what is a "normal" life?

Bob's lesson on perfection versus good came back to haunt me.

From another corner of the room Ricky quietly moved next to Alberto, who turned to speak to his baby brother. Before he could say anything, Ricky reached toward Alberto's face and, with a devilish smile, pinched Alberto's nose using his new thumb. Ricky squealed in delight at Alberto's expression, a mixture of astonishment and wounded pride. But there was a hint of pleasure also. This was a new skill, born of physical change. There was no stopping Ricky now.

Rosa leaned over to interject a comment. Before Alberto could intervene, Ricky pinched her nose as well. He laughed with obvious pleasure as Rosa made a funny face, tightly screwing her lips together and squinting. She didn't pull away. It was as if she wanted more, like exposing your face blissfully to a gentle rain.

I had lots of optimistic plans for Ricky. But, to be honest, I couldn't think of a better way to use a thumb, especially one like Ricky's, created out of string and baling wire—and love.

"It's imperfect, isn't it?" I asked Rosa.

"Absolutely." She nodded in affirmation.

"It's also perfect, isn't it?" I said.

"Without question," she replied. "How could it be otherwise?"

10.

Gossamer Wings

When I first met Kathy, she was a lithe, supple nineteen-year-old, gregarious, with curly brown hair, an appealing smile and boundless energy. The campus pizza parlor job was perfect for her. It gave her the opportunity to meet people in an informal atmosphere, and the salary and tips helped with school expenses.

The Friday evening I was called to see her she had been busy at work, catering to a massive, hungry college crowd. Kathy's job was to shred cheese for the pizza. She took whole wheels of provolone and muenster, cut them into blocks, and fed the blocks into a home-style grinder with a Teflon screw, the kind you use to extrude spaghetti or grind meat. The hopper opening was small, and a teflon plunger forced the cheese into the corkscrew gear.

Harried and rushed that night, she found it was faster to feed in the blocks by hand. To hell with the plunger. If the cheese blocks were stacked up, the one to be shredded was forced in by the one behind it.

When the paramedics brought Kathy to the emergency room, they brought in with them the disassembled grinder. In haste, she had pushed the cheese into the worm gear a little too vigorously and her small, delicate right hand was caught and pulled in. The photograph I took was of the corkscrew gear.

The Teflon corkscrew lay in a metal pan. Tightly wrapped around its threads was the deep flexor tendon to the long finger, violently pulled out of her forearm by the inexorable force of the worm gear. The finger, cold and lifeless, had been torn from the hand, the lower end of the tendon connected to bone, bits of muscle still clinging to the upper end. The other two fingers, also ground off at the base with their avulsed tendons attached, floated in a plastic bag next to the pan.

For her part, Kathy was Miss Cool—in public. From the moment it happened, though, I knew she knew this injury had forever changed her life. If she grieved over the loss, she must have done it in a big hurry, probably in the ambulance on the way to the hospital. By the time I first saw her, she was calm and lucid, and remained so throughout my treatment.

Jeff, her fiancé, wasn't.

"Surely there's something you can do to fix her hand like it was before. This is the age of medical miracles, for Christ's sake. I refuse to believe you can't put her fingers back on. There must be somebody around here who's competent enough to do it." His lips were drawn back in a grimace, exposing perfect teeth below a thin brown moustache.

"Jeff, don't be a schmuck," said Kathy. "Mom says Dr. Arem's the best. I want him to fix my hand, and that's that. What he says goes. If he says the fingers can't be replanted, they can't. End of story."

"Replace 'can't' with 'shouldn't,'" I said, adding a postscript for Jeff's benefit. "Eight laborious hours of microsurgery might stick the three central fingers back on and get them to live—barely. Maybe. They haven't been kept cold, the critical tissues are crushed to hell and they're badly contaminated. But, even if they survived, they wouldn't work worth a damn and they'd screw up the function of the small finger and thumb, all Kathy has left. An unacceptably long run for a painfully short slide."

"Yeah. Well, we'll see. I'm sorry. The future Mrs. Marshall simply isn't going to have finger amputations. If we go dancing. If we give a dinner party. Can you imagine the reaction? Endless questions. Endless explaining. Endless phony sympathy. I'm not sure I could deal with the revulsion, expressions of disgust hidden behind a caring facade. My dad knows Dr. Franklin. If he's available, let's find out what he says. He has a reputation for doing microsurgery."

Kathy's venomous glare was short-lived. The center of her makeshift bandage was staining bright red, the thick layers of gauze obscuring the amputations. She hadn't been moving. More likely, her anger raised her blood pressure enough to cause a little bleeding. It would quit soon. But the operating room was ready, and I needed to get her upstairs. "We have to go now, Kathy," I said. "Sorry, Jeff," she said, "I really don't want to wait for Dr. Franklin's opinion. I'll see you and Mom on the ward in a few hours."

Glancing back, I saw Jeff pacing the floor, wearing a sour expression.

Kathy was visibly shaken by her fiancé's reaction, relieved to be under way. I couldn't blame her. I, too, was disturbed by Jeff's obstinacy. I knew Franklin, knew what he'd say if Jeff reached him, so I wasn't concerned.

The anesthetic block was perfect, and I blessed my anesthesiologist for her skill. Superbly trained, she was refreshingly competent.

For incorrect reasons, finger amputations have been relegated, perhaps by default but largely by tradition, to inexperienced junior surgeons in training programs. I've always suspected this was because such programs in large universities tend to be frenetically busy, not outcome oriented, or simply understaffed. If you care about results, though, amputation is one of the most complicated and demanding operations in a hand surgeon's repertoire. Definitely not for amateurs.

The fingers work as a unit, referred to as a "community of digits." Not only are the tendons anatomically tied together in special ways, but the brain, the "master puppeteer," oversees their coordinated action. Deleting a part of the hand is tantamount to a functional lobotomy, and treating and overcoming the injury is like threading your way through a minefield, finding and defusing each mine as you go. Every tissue is important and must be handled thoughtfully to prevent problems.

Ragged bone ends must be smoothed, preserving critical length when possible. Managing joints, knowing when or whether to pre-serve them and how to do it, is a textbook in itself. Tendons are trimmed so they don't adhere to bone ends and get stuck in scar, lousing up movement of the other fingers. The major arteries must be ligated or coagulated to prevent bleeding. Each main digital nerve must be trimmed back in a special way to prevent formation of a neuroma (a tender lump of scar and regenerating nerve) in the amputation stump where it can easily be touched and drive people crazy with pain. The skin must be draped over the reconstruction with the finesse of an expert carpet installer adjusting to an uneven contour. Pull too loosely and you leave unsightly excesses and bumps. Pull too tightly, or take off too much, and you have inadequate coverage.

In the days before anesthesia, speed was the criterion of surgical skill. Especially in wartime, the best surgeons could remove a badly crushed, injured, or infected arm or leg in a few minutes. Of course, such rapidity in amputation did not permit the meticulous attention to detail required to achieve trouble-free healing. It's a miracle many of the amputation stumps healed at all.

The popular media image shows those in attendance getting the patient drunk in preparation for the surgery. Strong booze was relied on to provide more than mere distraction. It is not generally recognized that ethanol is a classic anesthetic agent; even complex abdominal surgery can be performed quite well under ethanol. The problem with routine clinical use of ethanol is that the anesthetic dose and the lethal dose are too close to make such use safe for patients. Undoubtedly, over the centuries, many were lost as this fine line was crossed, with fatal results.

Kathy's amputations were a typical mess. The grinder wasn't neat pulling off her fingers. The remaining central skin was ragged but sufficient to close the defect without loss of palm length and contour. The remaining small finger and thumb had superficial lacerations. They would heal promptly, without difficulty.

But this attractive young woman would live out her remaining years with a major hand deformity. Jeff's speculations were relevant. His attitudes may have been juvenile, but his observations weren't.

How would she deal with the losses, both functional and cosmetic? What about school and future employment? How much help would she need, physically and spiritually? How much emotional support, and who would provide it? Was she a sensible candidate for a prosthesis, and would this help her?

There was no saving the three central fingers. They were gone, never to return. The trick now was to make what was left work as well as possible. Kathy was way ahead of me.

"Dr. Arem, while you're there, can you do anything to strengthen my little finger and thumb? Spruce them up? Make me double jointed or something?"

"No, Kathy, I can't, but you can. As fast as the pain subsides, start moving like hell. Pull out all the stops. Don't let things stiffen. You'll make a lot of scar tissue—that's inevitable—but the scar needn't be tight or restrictive. You can remodel it to make it amazingly soft and loose. It's really up to you."

Silence. But I could almost hear the wheels in her brain churning furiously.

From beneath the surgical drapes came the sound of Kathy clearing her throat. "Doctor, it's me again. How long do I have to stay in the hospital?"

"Not long, Kathy," I said. "Just long enough to be sure your pain is controlled and there's no infection brewing. Maybe a day or so. Maybe less."

"I want to leave as soon as I can. But Jeff will want me to stay. He'll insist, I know."

"Up to you, Kathy. Do you have an ally, someone who sees things your way who can influence him?"

"My mom. She's a feisty lady. Been through three husbands, all dead now. Their doctors will say they had terminal illnesses, but I know better. Mom just wore them out. You'll like her."

"I do already. Anyone who can survive such a pile of misfortune and stay sane has earned my respect. At the very least, she'll have perspective."

"She also has a filthy mouth. I hope you're not easily offended."

I laughed. "There are few enough words in this language to say what you want to say. I believe in using everything you've got."

"She and Jeff don't get along too well. He has rules and he plays everything by the book. She doesn't."

"Kathy, I'm a prisoner of human biology. I don't have a magic wand. I didn't invent the limitations of the flesh I'm stuck with. Nature has a two-million-year head start on me, and I can't simply change those rules for anyone's convenience. Not Jeff's, not anyone's. Healing will progress at its own rate. In a young, normally nourished person like you—yes, you're well nourished in spite of the pizza—the rate is already optimal and no chemical, herbal or pharmaceutical, can speed it up."

I knew she'd like my conclusion. "Only your hand is sick. The rest of you is fine. I see no reason why you can't go home."

"You'll tell Jeff I can? He might listen to you."

"Only if he's not pissed off at me for not replanting your fingers. But I write the orders. If I kick you out, there's not much he can do. If you were married, he might have more of a say."

"Yes. Well, that remains to be seen."

Far be it from me to participate in a lovers' squabble. I finished applying the bandage, and a sobering bandage it was. There was no hiding the loss of the three missing central fingers. An optical illusion, I was convinced, exaggerated the visible space between the thumb at one end and the small finger at the other.

Kathy's mom was waiting for her.

"So. The master pizza chef returns, after a hard day in hell's kitchen. Now that you're finished rearranging the pieces of your hand, do you have any other recipes you want to try? Only, skip the ground meat. It's not healthy. Stick to vegetarian."

"Look Ma, I didn't plan this. It was just one of those things."

"Yeah, right. A crazy fling. But now that your trip to the moon is over, flit over here with your gossamer wings and give your mom a hug." Kathy smiled, tears welling up in her brown eyes, and grabbed her mother, careful to keep her bulky white bandage elevated, the missing fingers evident by the wide space between the small finger and thumb. Their hug was fiery, passionate, clutching. Desperate? Change is stressful, and both were dealing with enough tragedy to keep them busy for a while.

Jeff fidgeted in his chair, took shallow short breaths and looked pale. Excusing himself quietly, he fled to the room's bathroom and shut the door. Above the whoosh of the air conditioner I could hear the unmistakable sounds of retching and liquid dropping into the toilet.

Jeff, still pale, returned to his chair wordlessly. His discomfort wasn't lost on Kathy. "I'm sorry my bandaged hand bothers you so much. Don't I at least warrant a hug?"

"You and your mom seem to have everything under control. I didn't want to insinuate myself into your tender scene."

"Cut the bullshit, Jeff. Are you going to support my little girl in her time of need, or not?"

"Well, Verna, of course I am. It's just we all do it in different ways."

"Right now, I think Kathy would appreciate loving physical contact. You do believe in physical contact, don't you?"

"Look. I'm doing the best I can to deal with a complete assassination of my plans. This is a heavy blow. A heavy blow. You do understand."

"Do I? I understand you're a selfish asshole who doesn't give a piss for anyone else. Hard as it is, Kathy's going to deal with this, with or without your help. She's going to have my help, whether she asks for it or not. She'll learn new ways to do everything she did before. Knowing her, she'll amaze us. But then, she's an amazing young woman. I wonder if you appreciate what you're getting."

Jeff smiled a grim smile. But there was no warmth in it.

"Of course I do. And you're right. It will be hard." Jeff looked at me. Did I detect a sneer on his lips? "Incidentally, I called Franklin right away. He was noncommittal about the advisability of replantation. But he wasn't willing to come in, or to send in his team. So I guess we had an answer, of sorts." Jeff rubbed his eyes with the back of his hands. "Well, Doctor. What happens now?"

"Kathy stays overnight. Then she gets to go home."

"That's ridiculous. After all, she did undergo a major trauma."

"She's reliable. I trust her to take her temperature and report back to me. But there's no dead tissue left, she's on antibiotics, and the pain pills work fine, so I don't expect any problems."

Verna piped up. "Kathy will stay with me. We'll have a fine time."

Neither Kathy nor Jeff voiced an objection.

"Then it's settled," I said. "I'll write the orders. Verna, you'll bring Kathy in to see me for a dressing change next week?"

"Will do. In the meantime I intend to shower her with physical contact. I'm her mother, and rank hath its privilege."

As I left, Kathy and Jeff were looking warily at each other. From a distance.

The adjusters at the insurance company fawned over Kathy and, in an effort to minimize their liability, promised her the moon. They were willing, almost anxious, to send her to France that summer to have Jean Pillet, one of the world's most respected prosthetists, fabricate a cosmetically superb hand for her. Figuring her mom would be better help there than Jeff, she had no difficulty wangling a free ticket for Verna. The two extended their stay for a few weeks and had a grand time.

On their return, I got to see the prosthesis. As anticipated, it was a work of art. The texture was soft and resilient, fleshlike with color variations flawlessly matching her left hand. There was a second prosthesis—it is Pillet's custom to make a second, wintertime one, partly to use as a spare in case something happens to the first one, partly to account for seasonally induced fluctuations in hand skin color and complexion. The fingernails were superbly crafted and were made to accept a variety of polishes. The thumb web space, the wrinkles over the joints, the pattern of veins, all done with the sensitivity and skill of a Rodin.

Kathy showed it proudly to her friends, used it for a couple of weeks, then put it in storage and forgot about it.

It wasn't that she was ungrateful. Her lawyers worked out a generous settlement, including funds for her future education and vocational training, so she was comfortably set. It was just she wasn't the least bit self-conscious about her missing fingers. If anything, she was proud of what she had painstakingly learned to accomplish with what was left of her dominant right hand. And, as with everything, she improved with practice.

It never occurred to her to hide her hand. It was out there, for all to see. If you didn't like it—well, that was your problem, wasn't it?

Even Jean Pillet is bluntly realistic about his prosthesis. Motivation, he says, is "the attitude of mind which gives the amputee the will necessary to achieve a good functional result." A prosthesis should never be referred to as an "artificial hand." In his view it is not "a rival substitute for our marvelous hand." Even in this era of technological triumphs, a true hand substitute is beyond our reach. Luke Skywalker's new hand in *Star Wars* is pure science fiction. As Pillet says, "There is no miracle prosthesis."

Writing was a special challenge for Kathy. She supported the pen over the back of her small finger, holding it in place with pressure from her thumb. But the position was only for starters. Her graceful flowing penmanship was a source of special pride for her.

"It's all in the wrist," she told me with enthusiasm. "See? I can make a pen jump through hoops. You told me I could soften the scar if I really worked at it. So I did, and I did. At first my hand got tired, but it's toughened up a lot. I'm in school full-time, so I do a shitload of writing. Oops. I sound like my mom."

"How is Verna? I haven't seen her in a long time."

"She's fine. Just as profane and irreverent as always. I don't know where I'd be without her support. Making my hand work the way I wanted it to was damned hard. I've had my down moments, but she's such a positive person she always picked up my spirits. I couldn't give up."

"What about Jeff? I haven't heard you mention him."

"Jeff decided superficial appearances were more important to him than what's inside. We could never make it work. Not now.

He'll find a Valley girl to trot off into the sunset with. Me, I've got work to do. And I'm seeing other young men who are interested in *me*, in developing a sustaining relationship with *me*." Kathy threw up her arms, palms open. "I'm not helpless. I can really do quite a lot. I really can. I hold up my end as well as anyone." Red-faced, she hid the moisture forming in her soft eyes.

"Hey—I don't need convincing. I know what you've accomplished. I'm a believer."

"Don't get me confused with women in James Bond movies. I'm not into martial arts stuff. At least, not yet. But what's fair is fair. I don't expect any special treatment. Just equal treatment. Piss on employers who hire perfectly formed but functionally inept warm bodies. Give me a performance-oriented boss any day. I do my job. Period. Just give me a chance to show my stuff." Kathy jutted her chin upward and threw out her chest. She turned so I couldn't see her wipe her eyes.

How could I tell Kathy she was what made it all worthwhile?

11.

A Path Taken

Diseases that eat flesh have come to occupy a warm place in movie producers' hearts. The Ebola virus is a huge box office draw. It stimulates the imagination, gets top billing with no demands for residuals, and doesn't cost moviemakers a red cent.

But there is an ancient killer, humanity's scourge throughout recorded history, that stays in the shadows. It is responsible for more deaths than all the twentieth century's airline and auto crashes combined. It is always ready to change form, like a chameleon. In its atypical varieties, it ignores the best antibiotics, singly or in combination, that humans can throw at it. And it is magnificently capable of eating away bone and soft tissue despite a furious, and often desperate, treatment barrage.

Its names are legion. Scrofula. King's evil. White plague. Consumption. . . . Tuberculosis.

You might think that by the middle of the last century, like polio or smallpox, it was eradicated. You don't hear much about it.

Unfortunately, atypical TB is making a menacing comeback, to the consternation of infectious disease experts. It crops up under unusual circumstances and presents odd symptoms. Even with a good sample of infected tissue, initial stains may not show the TB bacterium, an innocuous-looking organism called a mycobacterium. Cultures may take as long as six weeks to grow and may prove negative, even as the infection is progressing, thumbing its nose at the treating physicians. The ramifications of a missed diagnosis are devastating, since valuable time—and tissue—are irretrievably lost. Not that treatment is so superb, even if the correct diagnosis is made. The drugs are toxic and, at best, usually fight the germ to a draw, keeping it from spreading but not eliminating it.

Mort Levy never wanted to be a teaching case. He wanted to do his job as a shop foreman, collect his paycheck, read his science fiction books, and have time to build his model ships before fatigue forced his eyes closed for the day.

The wood splinter that impaled his right palm in the work yard was no different from any other occupational hazard. A puncture that didn't even need stitches. A minimal event to be quickly forgotten.

So when his hand swelled, he went about his business and ignored it. Even when the swelling persisted for a week, then two weeks, he still ignored it. Hell, it didn't hurt. Probably the hand stayed swollen because he used it so hard at work. And his schooner in the antique bottle was almost completed. Just the mizzenmast and a few details to figure out.

The pain and redness were a rude shock. They upset his carefully organized schedule. Worse, they came under the scrutiny of his wife, Sarah, a former nurse's aide, who insisted he seek medical treatment. He grumbled, but he went.

The emergency room doctors decided he had an infection and called an infectious disease expert, who agreed. But infection with what? There was no open wound, no drainage, no pus. The offending contaminated wood fragment was long gone. Blood cultures were negative. Initial X rays were negative. The TB skin test was negative. So they became creative, seized on the pain as a major symptom, labeled him as a victim of routine "tendonitis" and started him on anti-inflammatory medicine. The aspirinlike qualities of the drug reduced the pain a bit, but did nothing else but upset his stomach. Now, his hand was becoming noticeably warm.

Rubor. Calor. Dolor. Tumor. Latin words for redness, heat, pain, and swelling. The four cardinal signs of inflammation. With nowhere else to turn, they called me.

Over the years, I've met a lot of neat people in crisis situations. But I would have loved starting my twenty-two-year friendship with Mort Levy in some other way. He is a bright, articulate, deeply spiritual man with strong positive attitudes (his salvation). He remained calm when, as Kipling said, those around him were "losing their heads." When I appeared, he viewed me with a detached, almost impartial interest combined with a resolute faith I would do my best in his behalf.

We introduced ourselves. "Call me Mort," he said. "We have what the NASA people love to call a 'situation.' I'll appreciate anything you can do to help, to get me unstuck. I have lots to do. I want to get back to work, but I can't with this thing the way it is."

His right hand was now massively swollen and red. The fingers hurt to bend. No one who had treated him had properly bandaged his hand to protect his joints against irreversible stiffness. No one had explained to him the need for elevating his hand, to reduce both the swelling and his pain. He was accurate in asserting he needed help.

"Mort, I agree with the diagnosis your physicians have made. It's about the only thing I agree with, but I agree. You have tendon inflammation, most likely due to an infection. Both for diagnosis and for definitive treatment, you need to have surgical exploration of your palm, to remove the bulk of inflamed tissue around the tendons, examine the tissue microscopically, and get cultures directly. Without this information, we're in a holding pattern and everything is sheer guesswork."

He processed the information wordlessly. "Doctor, let's get on with it. I have to do something. I can't stay where I am, and nothing has worked so far. I'm ready for some answers."

Was he? Was either of us prepared for what was to come? And would knowing have changed anything? I ask myself these questions, restless, tossing at night. There are no answers. It's why Robert Frost's "The Path Not Taken" is my favorite poem. When life's algorithm branches, we choose a direction, and there's no going back, no way to find out how it would have been different the other way, if it would have been "better" the other way. No way to relive it. No second chance. Things are the way they are. We learn value judgments are a waste of time; "better" and "worse" have no meaning. Choices have meaning. Contentment has meaning. Self-actualization and fulfillment have meaning. Living the adventure and feeling good about yourself, these have meaning. This is all there is, ultimately.

So I took Mort to surgery and operated on his right palm. I found visibly severe inflammation of the synovium, the tissue layer enveloping and lubricating the flexor tendons. It was red with a profusion of enlarged, engorged blood vessels. No localized tissue thickening into an inflamed mass (called a granuloma), no pus, nothing else. True tendonitis, but not at all typical, much worse than the garden variety we commonly see. I took lots of tissue for biopsy and for cultures, including TB and fungus cultures. Removing the relatively small amount of synovium I excised would have no adverse effects on his finger motion.

Microscopic examination revealed no bacteria and confirmed the diagnosis of inflammation. But there was something else, something peculiar. The tissue showed a feature called palisading, in which the thickened cellular layers, composed of special cells called histiocytes, were stacked up in a neat, ordered, and characteristic way.

Characteristic of rheumatoid disease, but he didn't have it.

Characteristic of fungus infections like Valley Fever, common here in Tucson. But he didn't have a fungus, either.

Characteristic of tuberculosis infection.

Previous cultures were negative. The TB skin test was negative. Was I missing something?

Unfortunately, we are reluctant, especially in medicine, to admit to the many limitations and imperfections staring us in the face, imperfections in diagnosis and treatment we live with every day. If only the stakes weren't so high.

At six weeks, then eight, then ten weeks, all cultures for mycobacteria were negative. During that time, Mort's hand didn't change. Swelling persisted despite elevation and a carefully applied hand dressing.

Here's where the algorithm split. The decision—the choice—was mine. Do I commit Mort to, probably, most of his remaining lifetime on harsh, potentially harmful or toxic chemicals based on tissue appearance alone, with negative cultures, a negative skin test, and a dubious history? Or do I wait, treat him symptomatically, protect his hand, prescribe work modifications and, maybe, go back later to re-biopsy and re-culture the tissue if symptoms don't reverse? Someone had to decide. Someone had to choose. And I was the "old maid," the last one in the chain of "experts," the one who, by default if nothing else, got to decide for Mort what was right. I mean correct. Dammit!

I elected the second option. In retrospect, I chose poorly. Over the next few months, infection continued and worsened. I re-explored his hand, and cultures, this time grew an atypical TB mycobacterium with the imposing name *Avium intracellulare*, a bad actor known to be resistant to almost all drugs and notoriously difficult to treat. With a sense of foreboding I started Mort on triple-drug therapy, based on the latest recommendations. Despite treatment, ominously, Mort's wrist X rays began to change. The wrist bones progressively dissolved and disappeared entirely, leaving his hand deviated, shortened, and articulating with his forearm bones.

I've never seen anything like this, before or since. Helpless is an inadequate descriptor of how I felt. Surgical removal of diseased tissue, whatever could be taken out without making his function worse, appeared to be the only option. I did this procedure on two occasions.

The medical jargon used to describe the setup we used in the operating room is "reverse precautions." This means not only is the field sterile to protect the patient from outside infection, but great care is taken to protect the surgeon and operating room staff from the patient's infection. The operating room staff were terrified, and vocal about it. The AIDS epidemic had started, and paranoia was rampant. Death or crippling disease were now significant risks to OR personnel, and they weren't happy about it.

"Dr. Arem, is this really necessary? Surely, there's some other way to confirm the diagnosis." The circulating nurse spoke for my scrub nurse, usually my friend, who simply glowered at me.

"Yes, it is. No, there isn't." I felt bad enough about the dramatic course this bizarre infection was taking. I didn't want to glorify it with a lot of talk. I didn't have to. Mr. Levy's X rays were hanging on the operating room view box. The change was unsettling. He had normal films when he first came to the hospital. Recent ones showed dissolution of his wrist. My OR crew routinely saw enough wrist X rays to appreciate the differences. They also knew his current infection was responsible for those differences.

All movements were deliberate, in slow motion. Almost fanatical steps were taken to ensure the removed contaminated tissue was properly sealed before it was sent to the lab for disposal. All sharp tools were handled as if they were "hot" radioactive isotopes. Paranoia over an inadvertent nick or scratch hung over the surgical team like a dark cloud. No one ever said "be careful." In the OR being careful was assumed. There were too many unknowns: How was the infection spread? Was it by droplets suspended in the air? Was it by contact with infected tissue? Was an open wound required? How virulent was it? Frightened, no one was taking any chances.

The appearance of the synovium encircling Mort's tendons and throughout his wrist wasn't sinister, any more than the appearance of a quicksand pool is sinister. What raises goose bumps is the knowledge of how deadly dangerous both are. The tissue was gray, swollen, and weepy with focal areas of yellowish thickening, highly abnormal. I could imagine the crypts and wet surfaces teeming with millions of hungry, aggressive bacteria. Part of me—the hysterical part—wanted to amputate the hand at mid-forearm and slide the whole thing into a plastic bag in preparation for incineration. Fortunately, the rational part, not the hysterical part, was conducting the operation. Carefully, very carefully, I sharply removed as much of the diseased tissue as I could, sealing the mess in a heavy plastic specimen container. There was no dotted line to guide me or tell me how much had to be left for function. A wrist fusion—or a fusion of what was left of his wrist—would have helped Mort's function, but doing it would have required widely opening and exposing the bones' inside surfaces. If massive TB infection invaded his bones through this route, I feared not only might we never control it, but that the infection, running rampant, might kill him.

It's customary, and good technique, to use copious irrigation with physiologic fluids to flush loose debris out of the wound. Normally such fluids, with some of the patient's blood, would soak into the large sterile paper drape sheets, to be collected and discarded. But this stuff? No one wanted to handle it. No one knew if it was *safe* to handle it. So I obtained a suction apparatus usually used by general surgeons for abdominal surgery. It has a plastic tip connected to a long length of tubing and empties into a sealed plastic container, hooked to wall suction. The entire unit is disposable. Small towels around the wound edges would soak up the overflow and also be discarded. I wondered if similar paranoia, similar fear, existed at Los Alamos when the first atomic bomb was being assembled.

The relief exuded by the OR staff after closure and bandaging were palpable. Eager to be out of there, they displayed an efficiency rarely seen in any bureaucratic institution. Mort was transported with measured haste to the patient holding area where his wife, Sarah, was waiting.

She was immediately confrontational.

"Well, this is a fine mess. A fine mess. What brought this evil to our doorstep? How, in God's name, can we cope with this?"

Answers. Everyone wanted answers. I wanted them, too. Stand in line.

"Morty is alive," I said. "Being alive creates possibilities. Nothing more. Nobody said anything about easy. If you want a walk in the park, climb in the carriage and hire a nanny to push it."

Why was I so feisty? Sarah and I were both angry, with anger born of concern, of helpless frustration. Mine was complex, in a different way. It was anger also born of guilt. Guilt over perceived inadequacies. Guilt over failure. I needed to wring my hands, both literally and figuratively. Justified or not, there it was.

"Look," I said, "I'm not the least bit happy about Mort's diagnosis. I feel betrayed by a capricious destiny. I feel toyed with, manipulated, jacked around. I've tried, without success, to gain the upper hand over this infection, and it's just laughed at me and mocked my efforts. At least we know now what we're dealing with. As far as I'm concerned, the buck stops here. Mort's going on triple-drug therapy. No more tissue loss, not if I can help it."

Mort and Sarah seemed placated by this reassurance. It was party-line stuff, and I said it with the finest intentions, but without confidence. Too many variables, many of them uncontrollable. In my twelve years as an active member of the Plastic Surgery Research Council, I had to design and carry out a new research project every other year to maintain my membership. My longevity in the organization was a reflection of learning how to conduct a study with only one independent variable and control the others. How simple it is to deal with merely one variable. If only real-life problems were as straightforward.

On his triple-drug regimen, Mort's infection stabilized, but at an additional price. His hearing deteriorated (was it the drugs?—no firm answer) and he became a proficient lip-reader to supplement his hearing aid. He developed back pain, severe enough to require narcotics for control, due to a focus of active TB in his lumbar spine. His orthopedic surgeons were unwilling to operate to remove this focus for fear of reactivating the infection. His right hand was a mess. When the TB gobbled up his wrist, it produced an inch and a half of shortening. This shortened the functional length of the tendons, robbing him of power and screwing up his ability to accurately grip or manipulate anything.

But despite his many infirmities and limitations, Mort continued to pursue his avocations, mainly building his ships-in-bottles. True, it took him five times longer to do things than it did previously, amid frustrations, inventing new tools to accomplish his ends, and muttered curses under his breath. But the quality of his modeling didn't suffer. If anything, it was better than before. *Inspirational* described both the man's work, and the man.

Years later, I sat facing him in my office exam room. "Mort," I said, "don't you ever get depressed?"

He smiled at me, a benign, placid smile. "Sure I do. But what's the point? We could have a pity party, and accomplish nothing. As Mr. Spock has said, it's 'illogical.' I'd rather put my faith and trust in a higher power, just like I have right along."

"Wait a minute," I objected. "You're one of the most rational people I've ever known. Where does a 'higher power' fit into the picture?"

Mort shifted his weight and looked down. His chronic illness had melted off over sixty pounds, and he hadn't even begun to gain it back. "My choice is not between rationality or spirituality, it's *both* at the same time. I don't pretend to have the answers. And I like the fact you don't pretend to have them, either. I'm a Reform Jew. My core religious experience is beyond logic and reason, and I don't have any problems with it. It doesn't mean it's irrational, only that the numinous transcends logic. There are deep mysteries here. They're not inexplicable, only I can't fathom the implications. But I keep trying."

PART II

An Informal History of the Hand

The history of the hand is the history of humankind. The patients' stories I've recounted, all based on actual cases, simply whet the appetite. What follows in this section expands on those stories and presents the history and basic information on each topic. Because humans are constantly at the fringe of new knowledge, uncertainty leads to disagreement and controversy. Wherever possible, I have tried to present all viewpoints without taking sides.

12.

Gesture

Nothing contains more expression and life than the movement of the hand. . . . The most expressive face remains unimportant without it.
—Gotthold Lessing

Gesture is a sign of character. This notion is as old as human culture. The ancient Chinese felt involuntary gestures revealed inner temperament. Members of the Bubi tribe in West Africa cannot communicate in the dark using speech because their language is dependent on gestures that must be seen to have meaning. In Brazil, the Puri tribe relies on gesture to make its spoken language comprehensible.

Gesture is a preverbal language. In infancy it is the only means of expression. Children convey what they want and feel through guttural vocalizations and body movements, This form of communication is entirely instinctual and emotional, yet the child is clear, in its repetition and insistence, about what it wants to communicate. The infant touches its face and its mouth, learning about distance and three-dimensionality through images supplied to its brain through its sense of touch. Its mind registers the internal brain-image of its mouth because its hands tell it about this body part.

Later, the acquisition of verbal language makes such gesticulation unnecessary for the expression of meaning. Yet the tendency to augment communication with hand gesture may, in some people, persist. Since gesture cannot accurately impart abstract thought, one may speculate that this superfluous movement is used either for added emphasis or represents an innate aspect of character development. This is certainly true for Angela, the Sicilian woman with carpal tunnel syndrome who was discussed in chapter 7.

Beginning in the 1930s, research restricted hand movements during speech to see if gestures assisted word production. Some studies found that suppressing motor movements of the hands resulted in decreased fluency, expressiveness, and intonation, increased hesitation, and decreased use of high-imagery words. However, use of iconic gestures, which represent imageable meanings of the verbal contents of words, facilitates list recall, augments story comprehension, and aids in the conceptualization of visual design. Thus there are no hard, fast rules. Kinesthetic individuals whose hands gesture constantly when they speak register great psychological discomfort if their hands are restricted.

It has been speculated that speaking a second language would entail a greater use of gesture to somehow make up for deficiency in verbal language proficiency or semantic content. In fact, studies showed the opposite was true. There was a positive correlation between verbal skill in the native language and gesture use.

Sir Richard Paget claimed, in explaining the origin of language, that we can make 750,000 gestures, far more than there are words in the largest English dictionary. Professor Alexander Johannesson of the University of Iceland found the origin of the sounds of language in the speech organs' imitating the sounds of nature.

Trappist monks, in deference to their strict rules of silence, developed a sophisticated sign language to allow silent communication. Traditionally, North American Indians communicated through sixty-five distinct linguistic families. Sign language and gesture were used when they met to barter, hunt, or decide on war. Hand configuration, body position, and movement were the classes of features that made up individual signs, usually with a rule-governed grammar. Even tribes with mutually unintelligible languages could communicate freely in this way. Northern and southern tribes had distinct dialect groups. But syntax was simple, succinct, and close to the images of nature.

The word *gesture* derives from two Latin roots: *gestus*, which describes the dramatic movements of the Roman orator or actor, and the verb *gerere*, which means to behave, comport, or show oneself. As opposed to posture, which is passive, gesture is active.

In *Gargantua and Pantagruel*, Rabelais played with gesture. His character Thaumastes (the name means "worker of wonders") challenged Pantagruel to a debate, and used a broad range of sign language in his arguments. Rabelais humorously spoofed the phony science and empty gestures characteristic of "scientific debates" of the time.

A common ritual is the handshake, considered in modern times to be a sign of friendliness. Paradoxically, it has also been a sign of mutual mistrust. In olden times, men would circle around each other, suspicious and prepared to fight. Before they sat down to talk, one would seize the other's hand to be sure he had no weapon. Psychologist Alfred Adler focused on the handshake and observed that the way people shake hands may reflect whether or not they are sociable and whether we choose to like or dislike them.

Crossing fingers for good luck reverts back to a belief in magic that can tie things together. Later, crossing fingers took on a religious connotation and became symbolic of the cross. Hand-clapping means applause in modern culture. It derives from the calling of spirits, to thank the spirit that performed so well for us.

Travelers who visit foreign countries and cannot speak the native language naturally revert to gesture to communicate. In such situations a well-developed kinesthetic consciousness is a real asset, and the language of gesture becomes an international idiom. Again, this is not a new concept. Under the emperor Nero, a governor who ruled over a population of mixed tribes on the edge of the Black Sea used an interpreter who was fluent with gestures to communicate with sailors calling at Italian and Greek ports.

Ethnic and cultural differences in the use of gesture in human groups tell the story of humanity. These have been chronicled in great detail by Desmond Morris. Meanings of gestures vary greatly from culture to culture. In our society, the "A-OK" sign, with the thumb and index fingers making a completed circle and the other fingers outstretched, is welcomed, an affirmation that all is well. In some Latin countries, however, it is an unspeakable insult. The "V for victory" sign, popularized by Winston Churchill, is correctly carried out with the palm facing forward. Reverse palm position, however, and make a V with the back of the hand facing forward, and the gesture, in some cultures, may have obscene sexual overtones.

Mediterranean people use hand gesture almost to excess. The history of Sicily is one of repeated uprisings and repressions, and perhaps gesture and sign language, combined with the people's emotionalism and temperament, emerged as a means of avoiding being overheard or understood by their oppressors. It has been suggested that the Sicilians, who lived in poverty for generations, kept themselves in a state of "voluntary agitation." This certainly describes Angela, whose wild gesticulations were, for her, a natural and unavoidable part of speech.

D. Effron, psychologist and linguistics expert, studied gestures among Italian immigrants and found that they "display gestures which extend away from their bodies . . . directionally, centrifugal movements." Again, Angela's circular arm movements fit well with this idea.

Polish psychologist Charlotte Wolff lived and worked in England, and pointed out, in 1945, that the restrained British used few gestures. But this feature, too, has fluctuated wildly over the centuries. The drawings of Rowlandson and, especially, Hogarth depict Englishmen of Shakespearean days gesturing wildly and with great temperament. A century later, Addison called gestures "unsuitable to the genius of the Englishman, even in rhetorical utterances." Still later, education produced the greatest restriction of hand movement, condemning gesticulation as undignified. Absence of hand movement in communication became associated with social stratification.

But to inhibit the free expression of inner dynamism is like damming a river. Subconscious motivations and reactions, in such circumstances, will seek a less obvious, more tortured outlet, sometimes with bizarre consequences. Slight movements of the hands, the face, the gait, retain an expressive quality, concentrated and revealing an existing undercurrent of emotion.

"Man without movement is 'dead,' " affirmed Dr. Wolff. "The vital concern for each individual is to find adequate expression for his inner dynamism."

In this regard, hands are indispensable for the performance and completeness of creative gestures. Think of all the varied means of human expression—whether in forms created by use of the hands, like graphic art, music, sculpture, or pottery, or in forms created by position and movement of the hands, as in dance, acting, or art forms in which hands are depicted. The posturing hand is an integral and essential part of artistic expression. The brain directs its effector organs, the hands, which shape and model the desired object or assume the desired pose.

John Wilson is professor of dance at the University of Arizona. Bald as Yul Brynner and graceful as Fred Astaire, he has long focused on and taught heiratic gestures, used in dance to convey deep inner or symbolic meaning. Different from the Stepanov Dance Notation or the newer Benesh Movement Notation—a movement alphabet used as a documentation of dance—heiratic gestures are extroverted movements often used as symbols of externalized conflicts.

An observation by the orator Quintilian in the first century presaged the development of sign language for the deaf. He wrote: "The hands not only assist the speaker but seem almost themselves to speak." The predecessor of sign language for the deaf, a one-handed finger-spelling technique, was developed in the eighteenth century by Abbe Charles Michel de L'Epee, later replaced by a two-hand method.

In 1815, Thomas Gallaudet, a young clergyman, teamed up in Paris with Laurent Clerc, a deaf French teacher, and returned to the first American School for the Deaf in Hartford, Connecticut, to teach sign language. Its successor languages are now in widespread use. Each country, each spoken language has its own sign language.

Today, American Sign Language, or ASL, is a fully functional language serving more than sixteen million deaf people in the United States alone. A language of hand gestures, ASL is in fact the native language of a half-million users in North America. It now has its own dictionary and technical manual and is prepared by deaf artists using models from the deaf community. Like many natural languages, it has regional variations. Gesture is a preverbal language used by all normal-hearing infants to communicate needs and desires. Deaf infants who do not vocalize "babble" with their hands, making random and progressively differentiated hand shapes, used later on in forming the signs of ASL. This babbling is carried out at around the same time in infant development as seen in children who use their voices to form their first words. When the hands of a deaf child must become the main means of expression, the child learns how with competence and at a typical age.

ASL uses hands to communicate with a combination of four gestural components: *location*—hand position relative to the body, fewer than twenty-five possible variants; *hand shape*—used in the formation of the sign, fewer than fifty possible variants; *movements*—hand motions used in executing the sign, twelve possible variants; and *orientation*—placement of the palms, twelve possible variants. Added to these are nonmanual clues like facial expression, body language, pauses, and use of three-dimensional space for grammar, all of which make ASL both complex and a true living language. Deaf children who learn only signing without spatial grammar wind up introducing such grammar into their signed utterances, even though they have never seen it.

In the last twenty years, numerous classification schemes for hand movement have been put forward. These largely deal with ergotic movements—hand usage and control tasks, manipulation of the environment—and with semiotic movements—communication of information. In any event, the close relationship between hand movement and language (communication) is one of the underpinnings of human development, separating us from other life-forms.

"The hands as tools of expression which convey the inner world of a person to the outer world are a human evolution," said Charlotte Wolff. "When man digressed from the animal through his upright walk the hands became free for new functions—one of them being the almost unlimited power of expressive movements. The language of gesture with which the hands are endowed is therefore one of the most valuable keys to the human mind."

13.

Sensibility and Touch

In the skin of the tips of our fingers we see the train of the wind. It shows us where the wind blew when our ancestors were created.

—Navajo, nineteenth century

We touch objects and thereby learn about our surroundings with our fingertips, with the highest density of nerve endings anywhere in our body except the tip of our tongue. Fingers interface with our environment and juggle it to suit our purpose. Archimedes said, "Give me a lever long enough and I will move the earth." The hands are the lever of the brain. With them, humanity has indeed moved the world.

As children, we learned there were five major senses: sight, sound, smell, taste, and touch. Blindness and deafness are painted as the most horrible losses, and the thought of perpetual darkness, of unbroken silence, of being cut off from physiologically normal humanity is immeasurably frightening. Yet somehow loss of touch is given short shrift. In our culture, touch is a neglected sense. Perhaps it is because our skin and mucous membranes are so richly supplied with nerve endings rendering their sensory territories sensitive to tactile stimuli. The skin arises from the same embryonic germinal layer as the nervous system. It is the largest sensory organ in the body (nearly twenty square feet, over 10 percent of body weight), and every square millimeter of it feels. So perhaps the notion of loss of feeling seems as remote, as unthinkable, as death.

Touch, based on unique sensory physiology, is referred to as the tactual-haptic system. Skin throughout the body is replete with tiny transducers converting all the sensations we feel to minute electric currents which are transmitted to the brain for interpretation, processing, and response. Think of the enormous variety of stimuli we experience: heat, cold, vibration, acceleration, rough texture, deep pressure, proprioception (position sense), touch perception, pinprick, and moving objects touching the skin, to name a few. Each microscopic transducer is named for the scientist who first discovered or described it.

But social contact depends on more than simple skin sensitivity. Consider the numerous variables involved in the impact and interpretation of social touch. Sensory mechanisms beyond the skin alone must be involved. Not to be left out are muscle spindles, cell and joint receptors, and a cellular workstation called the Golgi apparatus located in muscles, tendons, and joints. These anatomical features give the hand extraordinary versatility. It becomes necessary to divide touch into perceptual subsystems, as outlined by J. J. Gibson, an ecological psychologist at Cornell University. Skin and subcutaneous tissue can be stimulated without joint movement—*cutaneous touch*. If joints are also stimulated, the term is *haptic touch*. If skin and joints are stimulated in combination with muscular movement, this is *dynamic touch*. Skin stimulation combined with dilation or constriction of blood vessels is *touch temperature*. Inputs from skin and joints combined with balance receptors create *oriented touch*.

Of course, touching another's body, like all nonverbal behaviors, rarely has an absolute or incontrovertible meaning. A caress, a forceful shove, a light tap or a firm grip may carry a variety of possible interpretations. The significance of the contact depends on the particulars and peculiarities of the human interaction.

Jean-Paul Sartre said loving touch "incarnates the flesh." It makes us more alive, more aware of our physical existence. Our concept of self-image, of self-worth is tied to the messages we receive through touch.

The sense of touch is a wondrous gift. Our fingertips, with their exquisite sensitivity, not only contact and manipulate our environment but communicate with our fellow humans. They express love. It is a profound loss when a caress is a physical impossibility. How does one grope for substitutes? How does a still young, sensual woman, like Janice, Ted Abelson's wife, discussed in chapter 1, reconcile the living nightmare of a husband who breathes, who wants her, but can never again touch her as before?

Touch is basic, primitive, and arises early in life. A human embryo eight weeks after conception responds to tactile (*tact* comes from the Latin meaning "touch") stimuli, corresponding with maturation of critical portions of the nervous system. Throughout infancy, loving and supportive touch by a parent or caregiver seems to be crucial in promoting attachment and security, reassurance and self-esteem, safety and inner stability.

According to the thirteenth-century historian Salimbene, German Emperor Frederick II took babies from their mothers and reared them in isolation to find out what language they would speak without verbal stimulation. The babies all died, leading Salimbene to write, in 1248, that they "could not live without petting."

Data from the University of Miami's Touch Research Institute has repeatedly shown strong links between touch and healthy emotional development. Babies who are massaged, held, carried, or cuddled are less anxious and are less aggressive. In Bali, it is believed infants, when born, retain a God spirit. Consequently, they are held and carried everywhere and are not allowed to touch the ground for approximately eight months, when a special ceremony honors this event. Balinese infants are calm and cry very little. In this regard, their behavior mirrors that of infants in the Canadian Arctic's Netsilik Inuit group, where mothers carry infants on their backs for many months.

Human beings need touch, need physical contact with others. For this reason, solitary confinement is one of the worst punishments imaginable in a prison setting.

Anthropologist Ashley Montagu felt strongly that awareness was largely a matter of tactile experience. Failure to receive tactile stimulation in infancy "results in a critical failure to establish contact relations with others." Not surprisingly, interpretation of touch stimuli is culturally dependent. The way in which touch is administered produces, in human cultures, an enormous variety of responses and acceptable forms. Studies carried out by psychologist Harry Harlow in rhesus monkeys, however, focused on the importance of physical contact for normal social and, later, sexual development. Baby monkeys reared isolated from their real mothers vastly preferred contact with a terry-cloth covered "surrogate mother" over contact with a less cuddly wire mesh surrogate, ignoring food in order to experience the reassurance of touch. Harlow speculated that there is a "critical period" during which lack of such contact may lead to deep-seated behavioral deficits. Premature infants clinging to life in the sterile confines of an incubator are deprived of frequent human touch and cuddling and respond favorably to the texture of a sheepskin rug, with improved sleep, less agitation, and increased weight gain. Similar observations have been reinforced repeatedly by others in subsequent research.

Touch is healing both physically and psychologically. Patients with highly contagious illnesses treated in isolation and, therefore, deprived of skin contact by nursing personnel find the lack of human physical contact even more distressing than the illness itself. Many skin disorders have an allergic, immune, or neurological basis but nearly all are deeply affected by psychological state. If, by their appearance, they engender revulsion and avoidance in others, the subsequent depression may be significantly improved with touch. The offering by others of accepting, nurturing human contact goes a long way in affirming "I'm okay—even if my skin isn't." The same is true for healed, contracted third-degree burns, which may produce a grotesque, horrifying appearance, imprisoning the afflicted individual in what amounts to a natural solitary confinement.

We know our more expansive visual sense can be fooled more easily than touch, and lenses, lighting, and viewing conditions can distort perception. Our haptic sense is a more reliable indicator of reality and we "pinch" ourselves to confirm if we are awake or asleep. The plasticity of the nervous system, the ability to use whatever sense is working to substitute for deficiencies in others, is the basis for "sensory compensation," extraordinary skills developed by humans with sensory handicaps. Diderot, in his *Letters on the Blind* (1749), ascribed remarkable abilities to the sense of touch, and chronicled the high degree of tactile skill achieved by many congenitally blind people.

In this work, he presaged the later development of tactile reading in use today. Three-year-old Louis Braille, born near Paris in 1809, slipped holding a pruning knife while cutting a leather thong in his father's shop. The knife tip pierced one eye, severely injuring it. The injured eye was not removed immediately and became infected. Through a devastating complication called sympathetic ophthalmia, the infection spread via the optic nerve to the other eye, and young Louis became blind.

By age nine he was admitted to the National Institute for Blind Youth, whose founder had developed texts with large raised lettering. These were clumsy and inconvenient, but allowed the blind to read, albeit slowly.

Braille applied himself to his studies and became a professor at the school. While there, he witnessed a demonstration by Captain Charles Barbier de la Serre, an artillery officer who had invented a system of secret writing for use at night in covert military operations. Barbier's method, called "sonography," cut out code messages with a pen knife to create a twelve-dot cell, two dots wide by six dots high, each cell standing for a letter or phonetic sound, and decipherable in the dark by the fingertips.

While ingenious, Barbier's method embodied a flaw later clarified by modern physiologists. The spacing and functional skill of sensory receptors in the fingertips do not allow reading of a twelve-dot cell by palpation with a single fingertip touch.

But Braille quickly recognized the implications and potential and spent the next three years improving Barbier's method. Braille's major contribution was to reduce the size of the cell to six raised dots, each cell representing a letter of the alphabet. A six-dot code can be quickly "read" by fingertips passing lightly over the embossed pattern.

Like many new ideas, the Braille system met resistance and was not immediately accepted into pedagogical thinking. Critics pointed out that sighted teachers would not be able to read the books read by blind students. Critics also argued that reading raised print was more acceptable. Even though the experimental method was not widely used in 1869, William Wait, in simple experiments, showed that readers of embossed print were scarcely reading at all, while braille readers were fluent. Other arguments were equally silly and fortunately were not taken seriously by the blind. Opposition to the Braille system evaporated as it rapidly gained acceptance among the blind and their educators. Braille died of tuberculosis at age forty-six and did not live to enjoy the many improvements that were subsequently made to his basic design. One important modification has been the addition of contractions representing groups of letters or whole words commonly used in our language. This permits faster braille reading and reduces the size of braille books.

Despite the speed attainable by braille readers, haptic methods will always be slower than the visual reading of print. This reflects greater range and versatility of visual sensory inputs than of touch, more than it does differences in neural processing mechanisms.

For the blind, braille is the building block for language skills. It is a way of teaching spelling to blind children. Braille books range from mathematics to music, from modern fiction to law. Braille symbols appear next to the numbers in elevators. Braille is used for labels and for taking notes, and there are braille-adapted games, watches, playing cards, and thermometers.

Roland Galarneau, a visually impaired self-taught machinist, invented in 1970 the "converto-braille," a homemade electromechanical computer linked to a teletype machine. The original machine, improved through multiple generations of computers and advanced printing technology, forms the basis of high-speed braille printers used to produce braille texts circulated around the world.

Touch therapy, like massage, has been shown to decrease levels of cortisol, the stress hormone, and norepinephrine, the go-go neurotransmitter increased by stress and danger in the "fight-or-flight" response. Many human ailments have been shown to respond favorably to human touch. When used therapeutically, touch is love. There are strong links between this form of love and physical health and well-being. The "laying on of hands" has, both mechanically and via emotional intent, been a soothing and powerful healing tool throughout recorded history.

Considerably more controversial, however, is the modality known as Therapeutic Touch. Its underlying beliefs foray into ancient traditions and metaphysical realms.

In the early 1970s, Dolores Krieger, Ph.D., R.N., a nurse on the faculty of the New York University School of Nursing, and Dora Kunz, a self-declared clairvoyant and president of the Theosophical Society in America, developed and promoted the discipline of Therapeutic Touch (TT). Practitioners believe the human body possesses a "human energy field" (HEF), which extends beyond the skin surface. Illness is associated with, or even caused by, blockages, or knots of tense energy, in the flow of the HEF, diminishing its strength, and health suffers. Further, practitioners believe they can sense the HEF and, by moving their hands over the body's surface without physically touching it, can smooth out the energy flow, generate "impedance matching" in the tissues, and promote self-healing.

The practice of TT moves through five phases. First is centering, a form of meditation producing a shift in consciousness allowing detection of life energy and formulating an intent to help. Next comes assessment, in which the practitioner's *chakras* (nervous system regulators) act as nonphysical transformers or energy centers to convert the patient's HEF into physically detectable forms of energy, reduce pain, and induce relaxation. Imbalances are revealed and cues, insights, or intuitions culled, leading to an irresistible urge to help as the basis for treatment. The third phase, intervention or unruffling, rhythmically smooths out the HEF to repattern it in long, circular, sweeping strokes of the practitioner's hands two to four inches above the body, "decongesting" the accumulated energy and sweeping it down or off the feet, shaking the hands to purge them of excess energy. Simultaneously, the patient is encouraged to visualize healing. The fourth, or treatment phase, involves projecting healing energies through the hand chakras. The practitioner uses "effortless effort" to provide mindful action, send images of coolness to correct heat, and remain centered. The fifth and final phase is reached when the HEF is balanced and it's time to stop.

In yoga, the manifestation (from the Latin *manus*, or "hand") of self-realization, of union with the Divine, is a sensation of cool energy in the hands. Poet William Blake knew this when he wrote, in his epic poem on Milton: "With cold hand Urizen stoop'd down and took up water from the river Jordan, pouring onto Milton's brain the icy fluid from his broad cold palm." Many cultures have seen the hands as the portal to the state of the body and the nexus of psycho-spiritual transformation.

As in the "laying on of hands," healers going back to biblical times have promoted the healing powers of the hands. Ill patients in the Middle Ages sought the "royal touch" of kings to help and heal them. In the late eighteenth century, Franz Mesmer spread the belief he could heal through "animal magnetism" emanating from his hands, and that healing could occur only if the "subtle magnetic fluid" in the body were controlled or expelled. King Louis XVI appointed a Royal Investigating Committee, which included Benjamin Franklin and scientist Antoine Lavoisier, to study "mesmerism." The committee concluded that any cures effected were accomplished through the power of suggestion.

In eastern thought, the chakras (a Sanskrit word) are major nervous system regulators of the body's organs. If we are sufficiently integrated and sensitive, we feel the condition of our organs in our hands. Chakras are associated with specific colors, and Dr. Valerie Hunt at UCLA, working with Rosalyn Bruyere, claims to have measured the energetic frequencies corresponding to these chakras. In Muslim scriptures, at the time of the Last Judgment when Mahdi, the twelfth Imam, comes to save the world, we will set a seal on our mouths but our "hands will speak." Predictions by scholarly Muslims of the ninth century describe how Mahdi will integrate all religions by revealing the hidden unity behind them. Mahdi's companions (the realized souls) will receive instructions written by Mahdi on the palms of their hands.

China's *Tao Te Ching* asserts the right hand represents the principle of action, the left wisdom. In India, the left hand is associated with emotion and desire. Japanese, Tibetan, and Taoist teachings also echo these ideas.

Other systems of natural healing arose based on established principles of Asian medicine. Around 1900 Mikao Usui, born in Japan's Gifu district into a family of devout Tendai Buddhists, studied with Christian missionary doctors and was infatuated with Western science. Seven years after a near-death experience in a cholera epidemic, he came across a secondhand book unearthed in a Hokkaido monastery. The text revealed a complex tantra of spiritual practice, but it was incomplete, with major sections missing. In a search worthy of an Indiana Jones saga, Usui uncovered a second copy in Tibet, written in Sanskrit, and had it translated in Bombay and shipped to Japan. The two books together filled in the singular deficiencies of each. Usui was able to codify the teachings into a set of precepts, and created a system of symbols and affirmations using seven main hand positions. The precepts and symbols have been organized into a methodology known as Reiki (*rei* means universal and *ki* means life force). Similar to Therapeutic Touch, Reiki channels healing energy through the practitioner's hands to the recipient's body without actual contact. The Reiki practitioner uses intuition, sweeping his or her hands over the recipient's body looking for "hot spots" in the energy field which require healing energy.

Equally far-reaching is the healing system known as reflexology or zone therapy. Its origins may date back to ancient Egypt. Egyptian tomb drawings over four thousand years old depict hand massage. Modern North American Indians use this form of treatment, possibly deriving from Incan and Peruvian civilizations that flourished twelve thousand years ago. Cellini, the great sixteenth-century Florentine sculptor, is said to have applied strong pressure to his fingers to relieve pain.

Normally, a reflex is an involuntary muscle contraction produced by an external stimulus. In the context of reflexology, however, the term *reflex* connotes a mirror image or reflection. Reflexology maintains there are pressure points or zones in the hands and feet acting like small "mirrors" reflecting and representing every portion of the body. On the hands are prominent reflexes forming a miniature map of the abdomen, chest, neck and head, but all parts of the body are represented. Circulation of life force can become blocked, negatively affecting the entire organism. Intuitively, practitioners can detect such regions of blockage as "gritty areas" or "crystal deposits." Pressure applied to these spots in both the hands and the feet can release toxins, stimulate circulation, and dissipate energy blockages.

Nowhere are hands deemed more important than in the twenty-four hasta mudras, or hand positions. Mudras have been used for centuries by yoga masters to regulate energy flow through the body and, thereby, prevent illnesses that are believed to have their origins in imbalances in the chakras.

Mudras are essential to the interpretation of meaning in Hindu and Buddhist iconography. Buddha's hand is open to affirm he has no secrets from his disciples. Hand positions in Indian and Thai dance rituals communicate the meaning of complex sensitivities, transforming them into understandable signals. Echoing astrological and palmist ideas, the fingers and thumb are identified with the five elements: earth, air, fire, water, ether. Yogic tradition holds self-created chakra imbalances are the root of illness, influenced or cured by hand postures. Ayurvedic medicine is in line with these beliefs.

Controversy has raged with ferocious intensity over the validity of Therapeutic Touch. The concept of the "bioplasma" is not new. Indian spiritual traditions five thousand years old refer to a universal energy called *prana*, the source of life, sensed by religious adepts as light or energy around people's heads. The Chinese two thousand years later spoke of *chi*, universal energy divided into *yin* and *yang*, which could become imbalanced and cause disease. Kabbalah, the Jewish mystical philosophy that started in 538 B.C., referred to the same energies as astral light. Throughout the ensuing ages, artists have depicted light encircling the heads of Jesus and other spiritual figures. Later, Pythagoras also described such light. In the Middle Ages, Paracelsus called this energy "Illiaster." Its healing properties were echoed by mathematician Leibniz and hypnotist Mesmer. In the 1800s, Count Wilhelm von Teichenbach spent thirty years studying this "odic force," felt as being hot or cold by sensitive observers and conductible at low frequencies through a wire. In this century, physicians William Kilner and George De La Warr built instruments to detect this energy. Wilhelm Reich described it as "orgone." Drs. Lawrence and Phoebe Bendit extensively studied the "human energy field," and clairvoyants claim to have seen it. More recently, scientists in the United States, China, and Japan claim to have measured components of it.

To date, more than 100,000 people, including 50,000 nurses and health care professionals, have been trained in TT. Energy field disturbance is an official nursing diagnosis. Despite the widespread acceptance of TT, in 1996, fourth-grader Emily Rosa modified a science fair project allowing her to test the ability of twenty-one trained TT practitioners to feel Emily's energy field. All failed, and the results were subsequently published in 1998 in the *Journal of the American Medical Association*. Magician and skeptic James Randi has offered, so far without payout, over $1 million to anyone who could conclusively prove the existence of a human energy field. Opponents of TT have criticized a perceived paradigm shift into the healer's "intentionality," the use of "intuition" as a diagnostic tool, and other nondisprovable methods.

Rebutting these criticisms is Richard Lee's 1998 study purportedly demonstrating a change in the HEF by altering the hands' magnetic field. In general, the rebuttal argument would maintain that our technology is at the fringe of competence to measure energies currently at the threshold of detection. Our current inability to detect or measure psi, psionic, or paranormal powers like clairvoyance and telepathy may be in the same category. Perhaps the critical factors needed to settle the argument are simply the passage of time and improvements in technology.

One can convincingly argue that touch is the greatest sense in our body. The exquisite sensitivity of our fingertips gives us our knowledge of depth, thickness, and form. Major loss of skin is a death sentence. But with it we can survive the complete absence of any of the other senses. As Harlan Lane, professor of psychology at Northeastern University in Boston, has pointed out, if a sense is missing the brain sprouts new neural connections in the parts of the brain serving the residual senses. In effect, the brain reallocates to the remaining, intact senses brain areas that would otherwise have served the defunct senses. This remarkable brain plasticity creates the neural basis for enhanced performance, an adaptation favoring an organism's survival and function, even though sensory configuration has been altered.

Children who grow up without a sense of hearing use vision as a substitute. Children who grow up both deaf and blind channel all information through their tactile sense, primarily using their hands.

Normal language development occurs sequentially with nervous system maturation. Child psychologist Selma Fraiberg described how hands assume a midline orientation as the tonic neck reflex disappears, allowing the infant to gaze at its own hands and associate visual and tactile rewards with touching and grasping. By the end of the first year, eye-hand coordination has improved and the infant is pulled beyond its own body, gaining ideas about the world and its contents and exploring them. Pointing with the hands and looking to a parent for reassurance and affirmation are crucial preverbal gestures. In all children, evolving hand skills provide the foundation for language acquisition.

In a deaf-blind child, hands assume a pinching, biting, clawing posture as a sensory substitute for taking objects into the mouth. Lacking normal sensory clues to validate the permanence of objects and permanence of self, deaf-blind children take far longer than average to develop motor skills, and do so only with the intelligent, informed help of parents or other caregivers. The hands of such children must become curious, must learn to search and explore and to express a wide range of ideas, feelings, and emotions. Without education of the hands there can be no language, no cognitive development, no self-differentiation.

Human hands are a synthesis of powerful energies: the elemental forces; the centers of consciousness; the polarities of masculine and feminine. Hands control energy to heal, to transform, to balance, and communicate through touch. They interface with others. To be held by the Divine is to receive Divine spirit. Hence, the last words of Christ: *"In manus tuas Domine, commendo spiritum meum."*

14.

Left-Handedness

That raven on your left-hand oak
(Curse his ill-betiding croak)
Bodes me no good.
>   —John Gay

The great hypotheses of science are gifts carried by the left hand.
>   —Jerome Bruner

Was Stan, the biker whose snake injected its venom into his finger, sinister? Was he gauche? In the strictest linguistic sense, yes. Both *sinister* and *gauche* can mean "left," referring to Stan's left-handedness. The dictionary paints an unsavory picture of left-handers, with negative-connotation words like "defective" and "inept." The Italian word *mancino* means both "deceitful" and "left." The Spanish phrase *no ser zurdo* means "to be very clever," literally translated as "not to be left-handed." *Linkisch* is German for both "awkward" and "left." In Russian, *levja,* the word for left-hander, comes from the phrase *na levo,* which means sneaky. Even the Gypsies use the Romany word *bongo*, meaning left-handed, to refer to an evil person. Extending the concept further, let's look at the opposite. *Droit* means "right" in French, and it is no accident that the word *adroit*, with all its favorable implications, is lavished on right-handers. A right-hand man is a useful, trusted, and honored ally. Dexter, the opposite of sinister, refers to right. *Dexterous* means facile, useful, "handy" if you will. *Ambidextrous* literally means right-handed on both sides. *Right* is desirable. It means correct, and all that correctness implies. *Gauche*, on the other hand (no pun intended) means clumsy, uncouth, inappropriate in social interactions. *Left* derives from the old Saxon word *lyft*, meaning worthless. In Latin, *sinister* means left; it derives from the noun *sinistrum,* which means "evil." Wait a minute. How did we reach this impasse? What's the basis for denigration of left-handedness, for a stereotype amounting to vilification of southpaws?

Certainly nothing in evolution warrants it. Some animals, like cats and parrots, may show a preference for one extremity over the other. Peter MacNeilage and his colleagues have reviewed hand use in nonhuman primates and claim a consistent bias in favor of use of the left hand. But chirality (handedness) is a uniquely human attribute. Those who have studied prehistory and admired Cro-Magnon cave paintings thirty thousand years old have speculated that the animals were most likely painted by left-handed artists. This is because many are facing right, a pose not easily depicted, especially if done from life in a hurry, by a right-handed person. For the same reason, the ancient Egyptians were thought to be left-handed because, in their art, in their carvings, humans were often depicted in left profile, that is, facing right. The argument breaks down, however, when one notes these humans are holding and using objects with their right hands, suggesting right-handedness, in a percentage comparable to that seen in modern societies.

Despite claims supporting the existence of left-handed cultures, compelling evidence suggests "righties" flourished at the beginning of humankind. Countering the cave artist speculation is the large number of equally old paintings of hand outlines, made by holding one palm against the cave wall and applying paint with the other hand. With few exceptions, the outlined hand is the left one, indicating a creative—and dominant—right hand at work. Anthropologist Nicholas Toth examined stone flakes, produced when a stone core was struck with a hammerstone in the fabrication of early hominid tools, dating back a million and a half years, in Koobi Fora, Kenya. A preponderance show a clockwise flaking pattern. This suggests a right-handed toolmaker and can be distinguished from a counter-clockwise flaking pattern typically made by a left-handed toolmaker. Anthropologist L. H. Keeley, using Semenov's microwear analysis (a microscopic technique for examining the working edges of prehistoric tools) concluded, from rotary movements in wood bowls recovered from excavations in Claxton, England, approximately 200,000 years old, that the prehistoric woodworkers who created the bowls were predominantly right-handed.

Professor Chris McManus of the University College of London corroborates evidence of right-handedness in his studies of fossil records of Africa's Rift Valley which are over two million years old. Similarly, in the mid-1940s Raymond Dart found evidence that primitive humans of that era had used locally available natural implements as weapons to kill baboons. Examination of the baboon skulls revealed damage patterns in the front left area, as would be produced by a right-handed assailant facing its adversary. Further anthropologic evidence comes from examination of *homo erectus* and *homo habilis* skulls, looking at cranial morphology. Interior variations reflecting differences in the sizes of the two halves of the brain suggest a predilection for right-handedness. Other clues come from scratch marks found on fossilized human teeth. Early humans who ate meat held strips between their teeth, slicing the meat with sharpened stone knives (modern-day Inuit Indians from the far north use a similar technique). If the knife slipped, as it often did, it left scratch marks on the teeth. A left to right stroke pattern, made by right-handers, was found to be more common than the reverse.

Perhaps, one might think, there are some clues to the evolution of handedness in the long historical development of writing. The Phoenicians showed a propensity, in their writing and carving, for moving from right to left, a form known to be typical of Eastern as well as Middle Eastern cultures. Around 650 B.C. a writing style known as *boustrophedon*, meaning "ox-turn," evolved, in which writing moved in a straight line from right to left, then made a U-turn and ran from left to right. It may be wishful thinking to believe this seeming ambivalence actually marked a change in consciousness, a shift from a "mother Goddess," typified by a left-handed Amazon warrior woman paradigm to a "father God" mentality as humanity moved inexorably from a polytheistic to a monotheistic stance. Despite the right-to-left textual style, Michael Corballis, psychology professor at the University of Auckland, assures us that modern cultures that write from right to left are nevertheless predominantly right-handed.

As Mediterranean cultures expanded, their approach and philosophy, along with left-to-right writing, became the norm for Western societies down to the present day. A right-handed bias, based on its obvious success, was accompanied by widespread superstition coupled with the tenets and dogmas of organized monotheistic religions. Religions were tied to the state and, thereby, to daily custom. In all parts of the world, as monotheism gradually replaced polytheism, the operational paradigm was male-oriented. From Buddha to Mohammed to Christ, the right side was held to be valued. Even the Pythagorean Table of Opposites, recorded by Aristotle, associated the left with the female and with the dark and evil, while associating the right with the male and with goodness and light.

We thus see, in human history, an unfortunate and misguided attitude, a widespread cultural prejudice maintaining the right side is sacred and the left side profane. To be sure, this attitude, based on fear of the unknown and superstition, had existed centuries earlier. In ancient Egypt, the god Set (similar to Satan) was "evil" and "destructive" and was named "the Left Eye of the Sun." The god of life, Horus, was named "the Right Eye of the Sun." Ancient Mayan, Aztec, and Eskimo legends associated the left with sorcery and bad luck. The Maoris of New Zealand believe that strengthening and life-giving influences enter through the right side of the body, misery and death through the left. These convictions gradually infiltrated all cultures worldwide and were reinforced by the power of religious doctrine.

In Buddhism, the path on the left is the wrong way. The road on the right is the "eight-fold path" and leads to nirvana, or enlightenment. In Islamic countries, in India, and throughout Asia the left hand is "unclean" and is used for toilet necessities but *never* for eating.

The Bible contains well over a hundred references to the favorable attributes of the right hand. The left is never favored and is vilified a quarter as often as the right is praised. Satan sat on the "left side" of God, and was depicted as left-handed, baptizing followers with his left hand. In Jewish tradition, the Talmud describes the Chief of Satans as Samael, closely related to the Hebrew word for left, *se'mol*. The serpent who lured Eve into sin in the Garden of Eden was named Samael. In the Middle Ages, some people were tortured or burned at the stake because they were left-handed. A mole or blemish on the left side of the body was a mark of a wizard or witch. According to some theories, Joan of Arc was burned at the stake with this as an excuse. Wedding rings are worn on the left hand to ward off evil spirits. Courtroom oaths, shaking hands in greeting, the Pledge of Allegiance to the U.S. flag, and the sign of the cross are all carried out with the right hand. Nothing else is acceptable.

Many theories have been advanced to explain the complexities of hand-brain interactions. Some touch on the chemical foundations of life itself. Sugars are right-handed molecules, amino acids are left-handed, and DNA, the master molecule, is a right-handed double helix. Notions that these observations have any bearing on human handedness are fanciful and most likely not true. It was not until the modern era that science and study of brain mechanisms and anatomy permitted insights into the brain's role.

In its external appearance, the human brain looks a bit like a huge walnut and is bilaterally symmetrical. That is, both halves, or hemispheres, look the same. In nonhuman primates, the two brain hemispheres both look the same and, in the absence of language, work the same way. In most humans, though, the two hemispheres operate quite differently.

Hippocrates was the first to note a functional difference between the two sides of the brain. He observed soldiers with injuries to one side of the head who manifested disturbed movement on the opposite side of the body, and concluded, "The brain of man is double." It must have been jarring, then, when English physician A. L. Wigan autopsied a longtime friend and patient in 1844. Despite a lifetime of normal behavior, Wigan's friend, strangely, had a brain with only one cerebral hemisphere.

At around the same time, in the mid-1800s, Paul Broca, a French neurosurgeon and physical anthropologist, expanding the observations of French physician Marc Dax thirty years earlier, identified an area in the frontal lobe of the left hemisphere that plays a primary role in speech production. Shortly afterward, Carl Wernicke, a German neurologist, identified a portion of the left hemisphere deeper and farther back involved with language comprehension. Although there are exceptions, work by these and other researchers pegged language skills, a *functional specialization*, primarily to the left brain. Because motor fibers cross, the right side moves in response to left-brain instructions.

Roger Sperry's split-brain research in the 1950s studied people who, to prevent epileptic fits, had undergone severance of the corpus callosum, the major nerve bundle containing over two hundred million nerve fibers connecting the two hemispheres. With this enormous connection cut (by Pasadena, California, neurosurgeons Drs. Vogel and Bogen), the hemispheres could not communicate with each other, and each could be analyzed independently. Psychiatrist Juhn Wada at the University of British Columbia in Vancouver accomplished the same end nonsurgically, successfully treating epileptic patients. Wada dosed one of the two brain hemispheres by injecting sodium amytal through a carotid artery. Amytal anesthetizes and paralyzes each hemisphere independently of the other, allowing solo evaluation of the right or left hemisphere's control of speech.

Sperry discovered that the left brain concentrates on verbal and writing abilities, and is more linear, logical, intellectual, and factual, stressing cognitive activities. The left hemisphere is specialized to process, albeit almost instantaneously, one stimulus at a time. Like a computer, it is rule-governed, ordered, and sequential, organizing language into understandable syntactic units, recalling complex motor sequences, and is repetitively predictable. Sequential functioning produces linear behavior, linear thinking.

In contrast, the right brain is dreamier, more visionary, more intuitive, more holistic, and less pedagogical. It processes multiple complex bits of information simultaneously, in clusters. It is therefore capable of taking amoeba-like formlessness and transforming the unknown, the paradoxical, the novel, the ambiguous into recognizable patterns. The right hemisphere is essential to creativity. Impulsive, often unconscious, it excels at object recognition and visual spatial orientation as well as the processing of emotions.

For his groundbreaking work, Sperry won a Nobel Prize in 1981.

Stan Fotrell, a left-handed sheet metal worker, found acceptance in a shop in which getting the work done was paramount. Neither his colleagues nor members of his family knew him well enough, however, to comment on whether he displayed any of the presumed characteristics of right-brain–dominant left-handers.

State University of New York and University of Victoria, British Columbia researchers E. Goldberg and Louis Costa have pointed out that the brain's two halves are "wired" differently. The right hemisphere, with greater neuronal complexity and "interregional connectivity," specializes in tasks for which no preexisting brain routine is available but must be developed. The left hemisphere demonstrates superior ability to store and sequentially utilize routine preexisting codes and routines first established by the right hemisphere.

From a Darwinian point of view, there may have been a significant survival advantage for early humans to practice throwing and hunting maneuvers calculated and coordinated by the left brain and carried out by the right hand. Later, progressive reliance on language, writing, and toolmaking in the relentless advance of human culture may have been a strong catalyst for favoring right-sided motor abilities. Writing is a precise, detailed skill, an extension of language flowing from the left hemisphere out through the right hand. As humans became more settled, their tools reflected this predisposition. The scythe and sickle can be used only with the right hand. Family tools, usually right-hand oriented, essential for survival in all societies, were lovingly passed down to children, along with instructions in their use. If such use required primarily the right hand, this further encouraged the children to be right-handed.

Michael Corballis, in his book *The Lopsided Ape,* went so far as to attribute humans' unique abilities to a biological mechanism in the left hemisphere he termed the Generative Learning Device, or GAD. Corballis contended that the GAD enabled us to generate an almost limitless number of forms from a few elements, creating a basis for language, art, music, mathematics, and all elements of human culture. Right-handedness accompanied by language accounts for humanity's dominance over all other species.

If humans are genetically destined to be right-handed, then left-handedness represents failure to become right-handed. Why? In 1987, neurologists Geschwind, Behan, and Galaburda published a paper asserting right-handed children have developed normally, whereas left-handed children have suffered complications leading to anomalous cerebral dominance. According to their claims, genetics plays a relatively minor role. Chemical fluctuations or injury to the brain during development may alter cerebral dominance.

While in the womb both males and females share the same maternal placental hormones. The authors postulated that stress during pregnancy may lead to high testosterone levels. Coupled with testosterone coming from the developing testes in male embryos, the combination after birth retards left hemisphere development, leading to right hemisphere dominance and left-handedness. This would explain a higher incidence of left-handedness in males.

High testosterone levels inhibit thymus function in experimental animals both *in utero* and after birth. Since the thymus is an important link in immune maturation, the authors postulated thymus inhibition in left-handers would lead to abnormalities in immune function.

Furthermore, the authors hypothesized, if Paul Broca was correct and the left brain controlled speech, slowed left hemisphere development and excessive delays in speech controls would lead to learning disabilities and dyslexias, even though lateralization of brain dominance and speech production might be maintained.

Is all this true? Is alteration in the chemical environment of the fetus responsible for such a large panoply of alterations? The GBG theory, as the paper came to be known, has certainly "stirred the pot" and stimulated a great deal of research, some of which supports the theory and some of which shoots it down. The theory has not engendered complete confidence or overwhelming statistical validity. A sensible stance in the face of all this is, keep an open mind. Where there is controversy, uncertainty, and strong emotional bias it is wise to remain uncommitted and let the passage of more time—and human ingenuity—sort it out.

The GBG theory has been the subject of much debate. Offshoots of the theory have linked left-handedness with homosexuality, musical talent, AIDS susceptibility, and a tendency to die young. The last assertion has been the subject of numerous studies with no firm conclusions reached. From an evolutionary point of view, the GBG theory suggests a decreased survival fitness for left-handers. From this, one would expect the number of left-handers to decrease with time. Psychologist Stanley Coren, trusting in the inherent vanity of humanity, extrapolated handedness from figures pictured holding objects in ten thousand art works, dating from 15,000 B.C. to the present, and concluded from 1,180 unambiguous instances that the distribution of approximately 90 percent right-handedness has not changed for fifty generations. From this, it would seem either left-handedness has some benefit or at the very least is not hazardous enough to be exterminated.

Society has not made it easy for lefties. The litany of everyday devices favoring right-handed use is huge: corkscrews, construction screws, scissors, can openers, computer mice, most sports equipment, many musical instruments, and school desks to name only a few. In the industrial workplace, meat slicers, drill presses, band saws, and many other machines are designed for right-hand use. Left-handers have over a 50 percent greater likelihood of suffering injuries from these tools than righties. The ubiquitous right-handed bias in the physical world places left-handers at risk and creates the public perception they are clumsy and awkward.

As Stanley Coren and others have stressed, left-handers are still maligned and discriminated against. Coren said in 1992: "They may be one of the last unorganized minorities in our society, with no collective power and no real sense of common identity."

And yet, in their sheer magnitude, the contributions of left-handers to our culture are of inestimable value. From Babe Ruth to Michelangelo, from Marilyn Monroe to Leonardo DaVinci, from Albert Einstein to Queen Elizabeth II to Cary Grant to Paul McCartney—the list is not endless, but it's quite long. Although many left-handers excel at artistic or intuitive endeavors like music or chess, not all lefties are right-brain dominant. Just as there is variability and diversity in human trait expression, there is room for acceptance of such diversity which, ultimately, has only enriched our culture. Tolerance and restraint leading to enlightenment are in order. This is the true "right hand path."

15.

Intuition and Palmistry

There is nothing in the three worlds for knowledge besides the hand, which is given to mankind like a book to read.   —Hastha Sanjeevan

There have always been mysteries, gaps in our understanding, vast yawning breaches in the smooth fabric of what we know. Often, the greatest mysteries are close to us—or even part of us. Hands, with their complexity and their prodigious skill have, for millennia, been the object of human esteem and reverence. Tinged with awe is the recognition, becoming increasingly evident with our advancing knowledge, that hands are thought incarnate, the corporeal manifestation of human wish fulfillment. No wonder we have always held them in high honor, and want to know more.

Human handprints appear on the walls of Australian aboriginal caves, painted and preserved with naturally occurring materials like charcoal, lime, clay, and blood, an astonishing sixty thousand years ago. The handprints found in caves in Altamira at Santillana del Mar, Santander province, Spain; in caves in the Lascaux grotto in France's Dordogne region; and in Labastide in the French Pyrenees are more recent, a mere twenty thousand years old. Lacking scientific or anatomical knowledge, primitive people ascribed symbolic or religious significance to the lines of the palm. In a dangerous, often hostile world, how reassuring it must have been to draw significance and predictive wisdom from a body part, present yet unique in each individual.

Paralleling an oral tradition for construction of a god hierarchy, early humans had thousands of years to evolve an orderly tradition of palm line interpretation. Those who mastered the verbally transmitted palmistry framework, who could convincingly sell to their patrons the validity of the personalized content they saw or imagined in their clients' palms, became the great gurus and seers of the ancient world. Palmistry became a valued form of spiritual counseling. The advent of writing allowed a measure of consistency in their output and reproducibility in their schema of analysis. Writing, however, made a fortune-teller's job more difficult, since charisma and personal persuasiveness became far less pivotal in the client's acceptance of a palm reading.

Once humans got the hang of writing there was no stopping them. The earliest limited writing systems appear to have developed simultaneously in many parts of the world around the fourth millennium B.C. Pictures were still as important as they were in the cave painting days, and pictorial representation figured prominently in the output of early scribes. Hands had not lost their fascination. Sumerian clay tablets clearly show hands with a thumb and four fingers.

Desire to understand the lines in the palm remained strong. The origins of hand reading probably began in ancient India, before written history. In Indian tradition, the art of divination of both hands and feet—the Anga Vidya—was handed down by Samudra, the sea god. The Lord Krishna, an incarnation of Vishnu, was said to have special marks on his hands, foretelling greatness and mystery. Interpretation of lines in the hand probably began as a formal discipline with the Joshi caste in northwest India at least six thousand years ago. Their Sanskrit verses, or *slokas,* written in red herbal dye on human skin, were described by the noted palmist Cheiro (Count Louis Hamon). Jealously guarded by Brahmans in the cave temples of Hindustan, these writings contain the earliest known records of the long-standing belief that the hand's lines explain personality, which has come to be known as *cheiromancy,* from the Greek word *cheir,* meaning hand. The French word *chirurgien,* or surgeon, means one who heals with his hands. Both words come from Greek mythology, from the centaur Chiron, half man–half horse, the world's first great physician, teacher of Aesculapius and Hippocrates.

Parenthetically, Sanskrit was the original language of the Gypsies, nomadic people with roots in India who historically have been linked with palmistry, astrology, and fortune-telling, not always in a flattering way, for thousands of years. The Gypsies made palm reading glamorous for a time, particularly as they became prosperous in the fifteenth century. Gypsy fortune-telling smacked of clairvoyance, astrological mumbo jumbo enormously attractive to its receptive audience. Then it was revealed the Gypsies were spies for Sigismund, the Holy Roman Emperor, and they were treated with contempt and religious scorn. For centuries afterward, the art of palm reading was associated with the Gypsies and condemned.

To counter the Church's argument that palm reading was inspired by the devil, Gypsy fortune-tellers doing a hand reading argued that the devil shunned both silver and the sign of the cross. So, while doing readings, they frequently made the sign of the cross over a palm laced with silver. As reporter and palmist Elizabeth Squire has cynically, and probably correctly, assumed in her book on palmistry, the Gypsies kept the silver. Even today, in states like New York and Connecticut, laws make it illegal to accept payment for fortune-telling.

Cheiromancy came to be practiced in Egypt, Tibet, and China, but flourished in ancient Greece. Homer was reputed to have written a treatise, "On the Lines of the Hands." It did not survive, but was referred to by later authors. Pythagoras, in his book *Physiognomy and Palmistry*, traced the history of palmistry known, even then, to extend back thousands of years. The ancient Greeks, despite their sober, penetrating thinking, their belief in the laws of nature, their high level of intellect, were nevertheless tied to their magical ceremonies, sorcery, and superstition. Sickness to them meant being possessed by an alien spirit. Encounters with snakes, cats, or mice triggered superstitious beliefs and altered behavior, often for the remainder of the day. Astrology brought comfort, as well as terminology, and even today palm readers use the names of planets and of ancient Greek and Roman gods in their descriptions. Symptomatic of the times, many felt human fate was ruled by the stars. Such magical thinking was accompanied by a strong desire to decipher the lines of the palm and to read, in these lines, great destiny. Humans came to realize with certainty that their hands were the only possessions that distinguished them from all other living beings.

Anaxagoras taught and promoted cheiromancy, and wrote, in 420 B.C., "The superiority of man is owing to his hands." Pliny, Paracelsus, Plato, and Emperor Augustus thought highly of palm reading. Julius Caesar was an accomplished hand reader. According to fable, Aristotle, Alexander's tutor, found in a temple dedicated to Hermes a book on cheiromancy written in gold letters and sent it to Alexander the Great as "a study worthy of the attention of an elevated and inquiring mind." Aristotle believed one could identify a person's life span from the palm's lines. Later, around 350 B.C., he wrote a treatise on palm reading. Legend has it that when his palm was read for the first time his students were shocked by signs of faults and weaknesses there, but Aristotle accepted the faults as those he had struggled for a lifetime to overcome. To Aristotle, in his *De Coelo et Mundi Causa*, is attributed the quote: "The hand is the organ of organs, the active agent of the passive powers of the entire human system."

In pagan religions particularly, hands occupied a special position of reverence. Throughout prehistory, Walter Sorell, dance and culture writer, suggests, the magical power of touch in the laying on of hands was a common belief. But the juxtaposition of magic and miracles in the superstitious mind-set of humans was overwhelming. As monotheism became dominant over pagan beliefs, humanity embraced the almost miraculous achievements of the hands and associated them with divinity. The Bible (revised standard version) contains over twelve hundred references to the hand. The New Testament is filled with hand references. Since the Hebraic concept forbade any concrete representation of God, the hand, usually the right hand, was used to designate God's presence and actions. A hand emerging from a cloud is a common symbol of divine omnipotence. The cloud veiled the majesty no human could behold and live.

The Arab physician Avicenna included, in his eleventh-century *Canon of Medicine*, writings on the meaning of the hand's shape. After translation into Latin a hundred years later, this work became the basis for Europe's twelfth- and thirteenth-century enthusiasm for palmistry, a huge field of study in the Middle Ages called the *Samudrik Shasta* (Ocean of Knowledge). In this era, palmistry was part of the curriculum of major universities. The evolution of printing further fueled the fervor for hand line interpretation, from John Lyndgate's 1420 *Assembly of Gods Documents* (the first book on palmistry, then spelled "pawmestry") to Michael Scott's 1477 *De Physiognomia*, and beyond, to the present era.

In all religions, all cults, hands occupy a position of central importance. Although each finger has mystical symbolism, the thumb is the outstanding feature and, because of its dominance, denotes the godhead. This recurring theme signifies a recognition that hands are not only awe inspiring, a repository of mystery, but a source of great power. Hands became the facilitator, enabling translation of thought into reality, of imagination into creation. Hand development and brain development occurred together, exemplary of simultaneous evolution as both a physiological advance and a physical metaphor for the successful fusion of instinct and faith.

Polish psychologist Charlotte Wolff proposed evidence in 1943 that the brain's responses to many kinds of stimuli originated in the hand but spread to all parts of the brain. These responses, consisting of mental, nervous, and emotional reactions ultimately formed, after countless repetitions, what we call human personality. As noted, this was a basic tenet of cheiromancy, but Dr. Wolff gave it a thoughtful, psychological validation. In her book *The Human Hand*, Wolff stressed what she felt to be a fundamental truth: "The psychology of the hand is, like medicine, an art as well as a science. And, accordingly, intuition plays a part in it. But intuition must not be confused with clairvoyance. Intuition may be defined as the instantaneous synthesis below the level of consciousness of observed details, leading to the formation of judgments, only the results rising into consciousness. There is nothing supernatural about it." She maintained the subconscious had real existence and generated powerful thought processes. It rapidly formed subtle associations used by intuitive individuals with incisive effect.

Dr. Wolff was not alone in her musings. The great psychiatrist C. G. Jung wrote an introduction to Julius Speer's 1944 book on palmistry, *The Hands of Children*. In it, Jung commented, "Hands, whose shape and functioning are so intimately connected with the psyche, might provide revealing and ... interpretable expressions of psychical peculiarity, i.e., of the human character."

Dr. Wolff had a strong interest and background in cheirology, the scientific or psychological approach to "reading the palm." She, with her teachers Carl Gustav Carus and N. Vaschide, believed the hand's lines, shape, and configuration contributed to illness, physical constitution, temperament, and the interpretation, as well as the characteristics, of personality. Following investigations into palm markings in apes and monkeys, she designed methods for interpreting and understanding both constitutional and acquired tendencies in people. Her bias blended scientific information about the hand with a mixture of skepticism and mysticism. She viewed a legitimate cheirologist as one who read the hand with a scientific eye but spoke from intuition.

From their earliest days in medical school, all medical students are taught an important Latin dictum: *Primum non nocere.* First, do no harm. This vital teaching pervades the actions of every conscientious health practitioner. It is a lifetime injunction, taken seriously and with great solemnity. Since the mind has a unique ability to create a self-fulfilling prophecy of any prediction if it so chooses, any belief system that embodies some scientific basis with predictive power is potentially dangerous if used unscrupulously or indiscriminately on a naive or willing subject. When we start talking about brevity, or shortness, of a "life line," it's reminiscent of voodoo death—words *can* kill. It's probably just as well I never tried to read anyone's palm. I never took the enormous amount of time and effort to do it properly.

Wolff was among many to decry charlatanism and profit motive in any unsavory approach to palm reading. "One has only to mention the word 'palmistry' with the picture it calls up of gypsies at county fairs, to realize how exclusively the interpretation of the hand has become associated with charlatanism. . . . I have no hesitation in agreeing that cheiromancy, which is the superstitious reading of the hand lines, regarded as symbols of fate and character, does not possess a scientific foundation. Some of its traditions—those which can be explained on physiological grounds—are valuable. But its method as a whole, tending as it does to give psychological or mysterious meaning to every centimeter of the palm and every microscopic furrow, is an absurdity."

So how are we to reconcile a centuries-old practice, championed by great thinkers of the past, with our modern, rational, scientific viewpoint?

Dr. Wolff reminds us that occultism formed the basis of scientific discovery. Astronomy developed from astrology, chemistry from alchemy. Philosophy and psychology had roots in occult thought. There is a continuum connecting the archaic, in some ways quaint, notions of the past with the sophisticated present. It's a tempting mistake to disparage history's great thinkers simply for their lack of knowledge and technology—they did pretty well with what they had. And it's humbling to realize future generations will look back at our present era, with all its technical advances, as quaint and unsophisticated.

An anecdote, based on a true story, is useful for its perspective. An elderly woman blessed with a long, happy marriage had her palm read by a traditional cheiromancer according to a classification proposed by Elizabeth Squire, a folk hand reader (rather than an occultist or con-game artist). Shortly afterward, her beloved husband died. Consumed with grief, she declared life was not worth living, had no meaning or purpose, and gave up all her cherished activities, including bowling, needlepoint, and cooking. During this interval, a little over a year after her husband's death, the same cheiromancer again read her palm—observing that a number of the major crease lines were broken or partially obliterated, fuzzy and hard to read, and the "life line" was shortened.

Eight months later, the woman met a man, fell deeply in love, and completely reversed her negative attitude and behaviors. She resumed all her activities with a vengeance, added a few more, and declared she wanted to spend a long and happy life with her new partner. Her palm was read for a third time by the same person. Her crease lines were stronger and deeper, and her "life line" had actually elongated slightly.

Assuming the truth of the story, how does an objective person explain these events?

Following Vaschide's concept of an *image motorique*, Dr. Wolff would simply point out that a hand's movements determine its muscular development and, therefore, its flexion, or crease lines. Both the hand and a part of the brain's cortex retained imprints or "memories" of muscular patterns formed by countless movement repetitions—the *images motoriques*. When her husband died, the lady in the story markedly reduced her hand use and movements, altering the creases. Moreover, hand surgeons recognize that inactivity and lack of use often produce hand swelling, further obscuring palmar detail. When the lady again resumed her activities and increased the use of her hands, her swelling disappeared, hand muscle use resumed, and her palmar creases deepened. Cause and effect, with no "mystery" attached.

Of course, this is just one possible explanation of what happened.

Aristotle wrote: "The lines are not written into human hands without reason; they come from heavenly influences and man's own individuality." Palmist Johann Hagen (Indagine) wrote, in 1523: "It (is) madness to examine the hand and hastily interpret a life, and a body." And Walter Sorell wrote, in 1967: "The hand grasps—in the truest sense of the word—the world it creates . . . we shall always find that the greatest achievements in science occur simultaneously with an undiminished charlatanism.... Moreover, man, in moments of crises will always be caught reverting to the dark gods of magic and superstition. Despair and fear easily crack the thin veneer of this culture, and through the smallest crevice lurks the unknown, the primitive hope, the cheap belief."

16.

Phantom Limb

In 1551, Ambrose Pare, who poured boiling oil on wounds and later pronounced, "I dressed the wound; God healed it," wrote of the phantom limb: "Verily, it is a thing wondrous, strange and prodigious, and which will scarce be credited, unless by such as have seen with their eyes." He noted leg amputees often complained bitterly of pain in a phantom leg many months after the amputation. Those were tough times. Amputations were usually the consequence of war injuries and, if the limb wasn't cleanly hacked or blown off, were carried out unceremoniously without benefit of anesthesia, which hadn't been invented yet. Ominously, such amputees later came to describe their painful attacks as "nerve storms" often lasting days.

To his credit, Pare, one of the great physicians of his era, author of a monumental ten-volume surgical work, backed away from the use of boiling oil when he came to realize patients healed better without it, and earlier championed the use of the ligature, a surgical innovation in which sutures were used to tie off blood vessels to stop or prevent bleeding. While we find it intuitively obvious this would be advantageous, in the sixteenth century the method was controversial, almost heretical. Was such a lifesaving technique enough to offset the misery and human suffering produced by the boiling oil? Nobody kept score five hundred years ago, and we'll never know.

The concept of phantom limb, a sensory hallucination of an amputated body part, remained fresh in the public awareness despite a hiatus in reporting this phenomenon in medical journals. Admiral

Lord Nelson suffered from an amputated arm, whose sensations in his phantom fingers gave him "direct proof of the existence of the soul." Even the quintessential amputee, Captain Ahab in Melville's *Moby-Dick*, wondered, to the carpenter who was fashioning a new prosthesis for him, whether the newly made leg would finally "drive away" the lost flesh-and-blood one which "pricked" him at times.

"A person in this condition is haunted, as it were, by a constant or inconstant fractional phantom of so much of himself as has been lopped away," wrote American Civil War surgeon S. Weir Mitchell. "An unseen ghost of the lost part, and sometimes a presence made sorely inconvenient by the fact that while but faintly at times, it is at others called to his attention by the pains or irritations which it appears to suffer from a blow on the stump or a change in the weather."

Despite these clinical explanations, reassuring to us because they dispel the uneasiness of the unknown, our grasp of the phantom limb phenomenon remains elusive at best. The explanation, yet to be discovered, still resides in what Shakespeare called the "undiscovered country" and, like many things we dread, "puzzles the will." Although they may seem trivial, the observations that John Furman, described in chapter 4, still felt a ring on his phantom finger, or that horror movies made his phantom finger feel as if it was bending and straightening, hold keys that unlock the secrets of the mind-body connection. The conclusion, although a bit unnerving (a prophetic word) is that the mind and body *can't* be *dis*connected. The phantom limb manifestation originates in the brain, even though the injury is in the periphery. What was known to the mystics is only now gaining respect as nontraditional, alternative, or integrative medicine. The placebo effect—the ability to produce physiological improvement through mental intervention only—confounds the medical establishment. The brain's influence on the body, whether in modifying or even eliminating illness, creating a sense of well-being in the midst of physical chaos or in conjuring up a phantom, cannot be overstated. Placebo power makes it clear that alliance with, and encouragement of, the brain's overriding linkage with the body is far superior to pharmaceuticals as a healing force.

Early in the seventeenth century the great French philosopher René Descartes described a case of a girl who experienced phantom limb pain in her arm following removal of the arm at the elbow. He wrote, "Pain in the hand is not felt by the mind inasmuch as it is in the hand, but as it is in the brain." Such a mechanistic view was heretical in Descartes's time. Although he clearly understood that perception of skin events hinged on brain mechanisms rather than solely on sensory input alone, he had to be careful. Conventional wisdom at the time, heavily weighted by Church doctrine, held humans different from and superior to all other animals. Suggesting there might possibly be some overlap in brain methods for sensory processing was, quite literally, playing with fire. Descartes finessed the issue and appeased religious opinion by maintaining a *dualistic* view, distinguishing a nonmaterial soul from a material brain.

The mystery of the phantom limb phenomenon deserves deep thought and analysis simply because amputations occur so frequently. We are frail creatures existing in a harsh world, and a large cross-section of humanity has, in a lifetime, lost a limb or part of a limb. If the majority of these individuals experience phantom limb sensations, we are talking about a bizarre life event affecting a large number of people.

One of the most common causes for persistence of the phantom, often associated with persistent pain, is faulty amputation. The prime culprit in this category is the clear presence of tender neuromas, which form after traumatic amputations like John's, and are close enough to the body surface to be stimulated. When a peripheral nerve in an extremity is divided, it attempts to regenerate itself and grow back out to the tip of the limb. But the process, though biologically elegant, is imperfect. Scar tissue—"nature's glue"—always forms as a product of healing. It binds to the nerve and creates an impenetrable wall, a barrier to nerve regrowth. The scar, mixed with regenerating nerve, forms the tender mass called a neuroma. In a normal, healthy human a neuroma is invariably produced when a nerve is partially or totally severed. The messier the wound, the larger the scar and the larger—and usually more tender—the neuroma.

Because the transected nerve filaments are tangled in scar and don't reconnect to the microscopic transducers in the skin that register sensations such as hot and cold, vibration, touch, and all the specific sense modalities, tapping even lightly on the skin over the neuroma stimulates bare nerve endings and produces a sensation of intense, unpleasant tingling, like a painful electric shock. Neuromas don't disappear and generally keep pain alive as long as the patient is alive. Such pain is real, with a real physiological basis, not a phantom phenomenon.

On the other hand, phantom limb pain, pain felt in a lost limb, is puzzling. While nearly 100 percent of amputees experience phantom limb sensations, only about half experience actual pain. It's been thought neuromas are largely responsible for this, but two observations cast doubt on a definite linkage. The first is: pain may occur even though the amputation stump is free of sensitivity and tenderness, that is, there are no obvious neuromas. The second is: pain may develop before neuromas have had time to form.

So other causes must be responsible. For his part, Weir Mitchell thought pain was as likely due to irritation or inflammation (neuritis) in the nerves left in the limb. Certainly, a poor milieu in the stump—like infection, foreign bodies, or dead tissue, all unfavorable to prompt healing—sets the stage for the evolution of pain. Pain may be continuous or intermittent, may occur spontaneously or be induced by anxiety, emotional upset, damp, cold, or other stimuli, and may have a tremendous variation in the frequency, duration, and intensity of attacks. Victims describe the quality of their pain as knife-like, sharp, stabbing, gnawing, tearing, burning, or piercing.

One type of pain, however, demands closer scrutiny because it doesn't have a straightforward explanation. Some patients describe a severe cramping in their phantom arms, painful spasms not isolated to the point of amputation, which may last for hours. An intriguing theory maintains patients' brains are sending signals to muscles in the amputated extremity, muscles that are no longer actually there but which may exist in the phantom. Despite a true anatomic deficiency, the phantom limb is whipped into a frenzy of wishful movement by a brain that seems not to accept reality. Like a phonograph needle stuck in a groove, the brain repeatedly sends the same signal to move, over and over, to muscles that cannot respond properly because they're simply not there. Imaginary muscles, it would seem, are not immune to fatigue. Cramping and spasm, then, are due to overstimulation and overuse of phantom muscles.

Thus, can it be that phantom muscles are powered and driven by a delusional brain? This is a staggering conclusion which fits more in Lewis Carroll's *Alice in Wonderland* than the world of modern science. If you think things are sounding a little strange, let's pursue this line of reasoning a bit further, because it takes us to the dizzying edge of the connection between the brain and the hand.

In children age eight or older, virtually 100 percent have phantom sensations after amputations, particularly traumatic ones. But younger children generally don't, the percentages dropping off precipitously with progressively younger age. This intriguing finding at first suggests a gradual evolution and cementing of "body image" by age eight, strongly modified by sensory experiences and higher brain functions during development.

But there are cases of two-year-olds manifesting phantoms after limb amputation. Phantoms have occurred in even younger children, born without limbs. This observation suggests an intrinsic, inherited sense of self, of self as a whole person that resides in the brain, independent of everyday experience and hand activity. This innate, genetically predetermined self-awareness is the legitimate origin of one's body image.

What is the origin of the body-image concept? Wilder Penfield, the great neurobiologist, in a major breakthrough showed sensory input becomes organized in the cerebral cortex to form a miniature body model, a caricature with massive lips and a huge thumb, termed a homunculus. The body image as pictured by Penfield is a working image. The hands and lips, with the eyes, handle the enormous volume of sensory data crucial for infant development. Because of their importance, the hands and lips become incredibly detailed and functionally mature in the homunculus. Throughout life, the cerebral cortex accumulates and stores an ever-growing, complex array of experiences, fed by the hands with their vast powers of sensory discrimination, reinforced by repetition. By age six to eight, the basic body image becomes engraved in the brain.

But something is missing in this explanation. Such a notion of "body image" is an incomplete picture of the brain's schematic blueprint for body function. Penfield's maps of the sensory and motor cortex form only a crude guide to brain regions used to construct a stylized image of body-self. Losses involving these key areas of cortical surface may negatively impact, but not totally abolish, upper extremity functions. We all know that people who lose hand function adapt and "fill in" whatever is missing, using what works to make up for what doesn't. In this sense, such cortical areas described by Penfield are important but not indispensable. A more apt term is "necessary but not sufficient" for hands to work as they do. It is more appropriate to envision a vast neural network seamlessly integrating and managing the totality of bodily functions, inputs and outputs, sensory and motor requirements. Any part can fill in for any other part, albeit imperfectly.

The body operates a bit like a large organization. Upper management can type, send out bills, advertise, run machinery—in short, handle all the jobs required—but it's more efficient to divide the labor, assign specific jobs to individuals who can then hone their skills, persist, and become really *good* at those jobs. All in the interest of a smoothly running operation. Any part can do the work of any other part, but not as *well* as if the work is divided to allow specialization of task.

We are used to such specialization advancing human accomplishment to a razor edge. Watch an Olympic athlete or a concert musician if you need to be convinced that practice improves skill. Anything less is viewed as imperfect. In the human nervous system, how is such imperfection manifested?

When a skin nerve is cut, regrowth and regeneration are always imperfect. The result is abnormal feeling in the skin. Feeling is present, but somehow altered. How does one describe such alteration? Modern technology has provided a way of explaining the difference. Touching normal fingertip skin is like listening to music on a

CD. In contrast, touching the skin whose nerves had been severed and regrown is like listening to music on an AM radio. The essential *information* is there, but the quality is different, lacking in fullness, in richness. To use another analogy, a cartoon character is a sketchy representation of a person, recognizable as such, but certainly not as detailed, lifelike, or filled with nuance as a Rembrandt painting.

It seems as if each part of the brain contributes to the functioning of the entire individual. Each part carries within it a representation of the whole. In a pinch, each part can "fill in" for any other part, although imperfectly. This giddying thought suggests a holographic process operating. In a hologram, each part of the whole can re-create the whole. The greater the degree of fractionation, the smaller the pieces, the fuzzier the detail. This is true of the brain as well.

Think of the human body as a collection of many billions of nerve cells working simultaneously. Smooth, coordinated function of such a complex organism at all levels—sensory, motor, emotional, automatic, reflexive—depends on a well-orchestrated interplay of commands carried by nerves and executed by the muscles and tissues serving them. Analogously, think of the running of a corporation, the commanding of a military unit, or the performance of an orchestra, each with a *billion* players. Keeping everything straight would truly be a daunting task. The concept seems mind-boggling, yet is trivial in comparison to the complexity, the almost infinite connectivity, of the human nervous system.

Nerve cells in the brain can be visualized as treelike, branching almost limitlessly. This treelike branching has not been lost on scientists, who use the botanical term "arborization" to describe it. The small branches, called dendrites, are points of bridging, of connection (a connection is called a synapse) to the dendrites of other nerve cells. As Deepak Chopra and others have pointed out, there is now evidence showing that as we age new dendrites actually form, dispelling an old notion, taught to me in medical school, that adult nerve cells don't grow and new connections don't occur in the adult brain. Hogwash!

These concepts aren't easy, and many people have struggled

with them. But it gets hairier. To better grasp emerging concepts of nerve cell connectivity, Donald Hebb of McGill University described "cell assemblies," links of neurons, like dominoes, through synapses to interconnect widespread areas of the brain. Taking this one step further, Paul Grobstein developed the notion of "corollary discharge," an internal communication network to explain how one part of the nervous system sends information to another part.

In an even more elegant model, Canadian psychologist Ronald Melzack has described the neuromatrix, an overriding array of genetically programmed nerve cells ("upper management," if you will), distributed throughout the brain, which acts as the formative mold, processing and organizing huge volumes of data. This data, including reassuring feedback that the body is intact and one's own, is collated into a recognizable pattern, the "neurosignature," and sent on to the brain's central processing unit, the so-called SNH—Sentient Neural Hub—for its stamp of approval before being shipped out as an ever-changing stream of movement stimuli and awareness. Melzack, neurologist V. Ramachandran, Tim Pons, and others have in their research differed over how "cortical mapping," with territorial shifting and, possibly, new growth of overlapping nerve territories occurs. But the clear data that an infant born without a limb experiences phantom sensations in the missing limb is compelling evidence supporting a genetically determined neurological body acceptance.

Although highly technical, these models all represent a striving for basic understanding of how the brain works, based on observation of pathology in people.

Amputation of a limb is tantamount to setting off a large bomb in a crowd. What follows is pandemonium, confusion, disorganization, interruption of all normal means of communication, nourishment, and normal functioning. The more severe the mutilation, the worse and more long lasting the disruption. This interruption of sensory pathways between the limb and the cortex commonly results in a phantom limb phenomenon, in which the phantom survives as a hallucination of the body part which has been lost. This graphic metaphor also helps explain John Hughlings Jackson's century-old observation that the last position of the limb at the time of amputation is engraved on the cortex. A soldier in a foxhole whose leg is painfully twisted, then blown off may, sadly, experience a painful phantom echoing this uncomfortable posture for years afterward.

Enormous, virtually indescribable complexity is the hallmark of the human mind-body connection. The will to survive. The capacity for self-healing. Versatility and adaptability. All are characteristics of other animals but developed to a far greater degree in humans. I believe the phantom limb phenomenon represents an innate refusal to give up a functional body part intimately and irrevocably tied to consciousness. In contrast to internal organs and tissues that keep us alive but which are hidden, insensible, human limbs are indispensable to our psychic awareness. It is a fundamental part of our internal organization to view ourselves as having two arms, two legs, ten toes, and ten fingers. It should come as no surprise, therefore, that we hold on tightly but fail to let go lightly. When part (or all) of an extremity is gone, we try every trick in the book to live our lives and behave and function as if nothing had happened. Of course, this doesn't work. Nevertheless, we indulge in fantasy to maintain inner balance, a reassurance that all is OK. Inner acceptance of the phantom requires what has been called a "willing suspension of disbelief." Phantom limb is a mind game, mute testimony to the overwhelming intricacy of the nervous system.

Since it is nerves that carry messages throughout the body, attention to the nerves is paramount in any amputation. John Furman would have benefited if more care had been taken in correctly treating the nerves in his finger amputation. There are only two proper ways to treat acute nerve division and prevent pain. Normally, one way is to achieve a near-perfect nerve repair so the severed nerve filaments connect to their mates on the other side of the gap, restoring sensibility and motor function. Once nerve regeneration has concluded, neuroma symptoms will abate. This is impossible in an amputation, since the nerve on the amputated side is irretrievably gone, along with the rest of the amputated tissue, and attempts by the nerve to reconnect remain futile and "unsatisfied." The trick, then, the second way, is to trim the divided nerves back to a level where they are covered by ample soft tissue so when the expected neuromas do form, they are protected, removed and distanced from the constant bombardment of mechanical stimulation. If you can't tap or press on a neuroma, it is usually symptom-free. Usually.

Unfortunately, surgical training programs often perceive amputation, especially of a finger, to be trivial and assign an intern or junior resident to handle it. This is a big mistake. The average person dealing with the major psychological blow of a missing body part usually doesn't know enough to insist on having a senior surgeon handle it. Proper surgical closure of amputation wounds is one of the most complex and demanding tasks in hand surgery. Skin flaps must be preserved, rotated, trimmed, and sutured gently to keep them alive and provide adequate closure. The bone end must be trimmed and smoothed, avoiding shortening the finger too much but being sure there is enough skin and soft tissue to cover the bone end without tension. Blood supply and tissue survival must be assessed so only living tissue is used for closure. A complex protocol guides the proper handling of tendons, ligaments, joints, and, especially, nerves. Doing all this well requires experience and expertise. Doing it poorly leaves patients with residual problems. Among the most serious are tender neuromas, like John's, with persistent phantom pain. Most patients with amputations develop phantom sensations. But pain is another story. Long-lasting pain in amputation stumps is preventable.

My friend Phil is a perfect example of what can go wrong. Phil lost his left leg at mid-calf in a school accident when he was fifteen. Luckily he had a good surgeon who provided him with a well-nourished, healthy amputation stump leading to a smooth healing process with no infection or complications. The nerves were covered with ample thickness of skin and soft tissue, which situated them well away from the wounds and incisions. He was fitted with serviceable below-knee and foot prostheses and learned how to walk with them. Phil experienced typical phantom limb sensations initially but, after a year or so, they diminished and, ultimately, disappeared.

But twelve years after his injury, at age twenty-seven, Phil lost a lot of weight, his stump shrank a bit, and his prosthetic fit became a little too loose. Naturally skeptical and cautious after years of lifestyle modification, he was initially reluctant to do anything. But his doctors were reassuring and convincing. "What could go wrong?" they said. Phil now curses the day he was talked into undergoing a stump revision. His surgeon—a new one, not the one who did his original amputation—freed the major nerves and pulled them down close to the incision, leaving them more exposed with less coverage than before and, probably, handled them roughly.

Phil's phantom sensations returned with a vengeance, this time accompanied by burning pain, which, to this day, thirty-one years later, has not let up and which has never responded to any form of treatment. And he's had lots of treatment. More than forty methods have been proposed (and used) to treat phantom pain, ranging from massive narcotics to implantation of an electronic spinal column stimulator. Most of them have been tried on Phil, to no avail.

Interestingly, one approach has aided Phil more than anything else. Distraction. If he is able to focus intently on some activity or preoccupation that pulls his consciousness away from the experience of pain, the pain lessens. Positive thinking, creative visualization, and mood enhancement all also seem to help. This observation supports Ronald Melzack's neuromatrix theory, as well as his idea that the limbic system, the part of the brain that oversees emotions, contributes to the perception of phantom pain. As Milton wrote in *Paradise Lost*: "The mind is its own place, and in itself can make a heaven of hell, a hell of heaven."

17.

Skin Grafting

The earliest skin grafts were animal skin (xenograft) as a burn dressing. Frog skin, still in use today in Brazil, was used to cover burns fifteen hundred years ago. Later, in the sixteenth and seventeenth centuries, reptile or lizard skin enjoyed some popularity. But pigskin was most similar in its characteristics to human skin and was relatively cheap and plentiful. In the absence of anything else, pigskin, introduced clinically in the 1960s, was in great vogue for a time and is still the most commonly used xenograft burn dressing. A xenograft's limitations are that it cannot develop blood vessel connections with the burned host and, coming from a different species, will reject and slough off. So it is certainly not a permanent fix.

First introduced for clinical use in the 1960s, human skin grafts harvested from deceased individuals, preserved cold in glycerol, and stored in liquid nitrogen vapor seemed ideal. Such grafts could be procured in abundant supply, would "take" like the burned patient's own skin, and, being human skin (allograft), would potentially look normal and function normally.

In major burns, there is no question that human cadaver skin can be lifesaving as a temporary measure. There are, however, three significant stumbling blocks to more widespread use of allografts for permanent burn healing. The first is availability. I've worked in a burn unit as an intern and skinned a cadaver with a dermatome. It's a lot of work. Freshly harvested cadaver skin works better with better viability than stored, preserved skin, and there is simply not enough to go around. The second is potential transmission of diseases, like cancer or AIDS or other viruses. Despite screening guidelines from the FDA and the American Association of Tissue Banks, this is still an incompletely solved problem. The third is immunologic rejection. We are in an era in which many different organs—most recently human arms—can be successfully transplanted from one individual to another, and kept from rejecting with powerful anti-rejection drugs. But skin is, immunologically, quite active, and drugs to prevent rejection, with their toxicities and side effects, would probably be needed for a lifetime. For a small burn, with plenty of the patient's own skin available, allograft application is a long run for a short slide.

Experiments have shown cultured skin cells are capable of regenerating stable, normal-appearing connective tissue and an epithelium, or surface layer, indistinguishable from normal skin. High costs and disappointing early results have shelved this approach for now, but there is hope for the future.

On the other hand, if real human skin can't logistically be applied to the burn area as a permanent solution, what about synthetic products? The exciting notion of applying synthetically fabricated skin substitutes to fresh burns has spawned an entire industry.

Unfortunately for Sam, in chapter 5, none of these new products had been developed or was available at the time she burned her hand with flaming grease. Now available are Biobrane, Transcyte, and Integra, all membrane materials containing cells and supporting matrices. All work well to temporarily resurface burn wounds. And new ones with superior properties and greater likelihood of permanence are being developed all the time.

18.

Carpal Tunnel Syndrome

Every era has its disease: at the beginning of the last century, melancholia and tuberculosis, at its end, HIV and AIDS. This decade, fueled by technology and enormous productivity demands, will go down as the age of cumulative trauma (repetitive strain injury) whose centerpiece is carpal tunnel syndrome, or CTS.

CTS is increasing to monstrous proportions in the workplace. Bureau of Labor statistics show that low back injuries are declining but the incidence of CTS is rising, with over two million new cases annually and no end in sight.

Eighteen years ago I consulted for a manufacturing firm dealing with an increasing problem with CTS. When I analyzed the production line, the reasons were clear-cut and I made recommendations for changes that entailed a high up-front cost but would radically reduce the problem. I was told the following: "When the cost of medical care exceeds the cost of making the changes, we'll make the changes—but not before."

Such a callous attitude may have been an affordable luxury in the early 1980s, but it isn't now. The price tag for surgical and rehabilitative treatment of one CTS-afflicted hand, including time lost from work, is $30,000. Both fiscal and humanitarian responsibility demand solutions.

Unfortunately, solutions call for grudging acquiescence to the frailties of the flesh and an understanding of human biology, which has a two million year head start on us. CTS in the workplace represents a failure of tissue adaptation, a failure of tissues to compensate for excessive demands placed on them. Why should this come as any surprise? All things fail if they are overtaxed. Appliances, furniture, and autos wear out and break down if pushed too hard. Why not the human arm?

The word *carpal* means wrist. There are eight small carpal bones that link the forearm to the hand and fingers. Running from muscles in the forearm are nine flexor tendons, which attach to the bones of the fingers. When we want to squeeze, pinch, lift, write, type, or do just about any activity, the finger flexor muscles contract with power, pulling on the tendons that bend the fingers.

To gain access from the forearm to the hand, the tendons run across a U-shaped channel—the carpal canal—a conduit formed by the special configuration of the carpal bones.

When we hold or pinch strongly there is a natural vector of force wanting to make the tendons lift up and bowstring out of the carpal canal, robbing us of grip strength. To prevent this, nature engineered a thick, tough ligamentlike structure arching over the roof of the canal and sealing it off, making a closed, tight space only about an inch across—the carpal tunnel.

Where the tendons run through the tunnel, friction from movement would wear them out. To prevent this, the tendons, each the diameter of a flattened pencil, are enveloped by a filmy, delicate layer of tissue, the synovium, which actually manufactures a high-grade oil. Under normal circumstances, this self-lubricating system works exceedingly well.

In an arguably poor piece of celestial engineering, in addition to the nine tendons nature also placed inside the carpal tunnel a nerve about the diameter of a pen, the median nerve, which carries to the brain critical sensory information from the thumb, index, long, and thumb side of the ring fingers. There is barely enough room in the tunnel for its normal contents. If a pathologic process like swelling stuffs into the carpal tunnel material that doesn't belong there, it's like overstuffing a pressure cooker. The walls are rigid and unyielding, and the only thing that can happen is the pressure goes up.

The median nerve is exquisitely sensitive to pressure. Any increase in pressure chokes off the nerve's circulation. When this occurs, numbness develops in the affected fingers. This numbness is what is known as carpal tunnel syndrome. It is important to note that pain is *not* a characteristic of CTS. Pain accompanying CTS usually signifies something going on apart from nerve compression.

In the typical modern workplace there is a lot of work to be done. With automation and shifts in population, a greater burden is placed on fewer people. Motivating forces are strong: fear (of being fired or reprimanded by supervisors); the lure of more money; hype about greater productivity and the success of the business. These forces translate into increased output, often beyond the limits of what the tissues of the upper extremity can comfortably tolerate.

Have you ever sprayed weeds with a spray bottle? Money out of my pocket says you can't spray continuously with one finger for more than one minute before you tire, experience forearm pain, and have to change fingers, change hands, or stop. Why? Because the flexor tendons and muscles are not built for continuous use. When used to excess, the tendons alter their physical configuration and require time afterward to recover. If recovery or rest time is not provided, a process of chronic irritation begins. Bioengineering studies of soft tissues have characterized these changes, but the precise parameters of strain and rest in humans are not known.

People have been using typewriters for years without CTS developing. Why, in this age of computers, is it such a problem? There are many potential explanations, including poor ergonomics and physical setup in the typical workstation and ceaseless workloads. But one simple answer may be the absence of a carriage return. With all mechanical typewriters, even with the Selectrics, there is a momentary pause in the repetitive finger activity of typing as the carriage return brings you to the next line. These momentary pauses cumulatively add up, and may actually be critical in the biomechanical recovery of flexor tendon systems. With the computer, which advances you to the next line automatically, you never stop. If you add to this mandatory line counts and incentive bonuses to encourage you to type more than your tissues can comfortably tolerate, you simply put up with the discomfort and keep going.

This is a crucial point. The wrist/forearm pain that usually begins the whole process is a sign of overuse. It is not in itself CTS. The analogy of the pressure cooker may be a good one in this context. When you begin to heat up vegetables, it takes a little time before steam is generated and the little rocker on top begins to shake. With tendon overuse, it takes a little while before synovial swelling builds up sufficient pressure within the narrow confines of the carpal tunnel to squeeze off circulation to the nerve and produce numbness. So there is often a delay between the onset of pain and the onset of numbness. Diagnosis of tendon irritation, by physically placing the tendons on passive stretch, is the "smoking gun" linking the work (or injury) to the evolution of nerve compression.

When the roof of the carpal tunnel is opened surgically, the nerve is immediately decompressed and the numbness quickly abates. But the tendons, chronically deformed by weeks or months of overactivity, take much longer to recover. Pain with hand use persists, even though the numbness is gone. If a worker with CTS caused by tendon irritation goes back to unrestricted work too soon after surgery it is persistent pain, not numbness, that forces a cessation of work. For this reason, nerve decompression using a special endoscope may be doomed to failure. While skin healing with small incisions may be more rapid, sadly, despite the hype, visibility and exposure are poor and it is not possible to see or adequately treat the tendons. If the CTS is due to tendon overuse (as is commonly the case), a rapid return to work may be ill advised. One of the potential advantages of open surgery is that in addition to providing better and safer exposure of the nerve and tendons, a longer recovery provides an excuse to procrastinate before returning an employee to the workplace. Often, this extra time allows the tendons to recover sufficiently to do the work required of them.

Clinicians have termed the pain tendonitis, but this is a misnomer. If one biopsies the synovial tissue covering the tendons it may be boggy, swollen, thickened, or irritated, but not truly inflamed.

There is no "itis." The whole process may be reversible if one begins remediation quickly enough.

Many people afflicted with these nerve and tendon problems experience pain in their shoulders, neck, and back. A major reason for this is what I call the "rock in your shoe" phenomenon. If you get a rock in your shoe, painful to step on, and you don't stop to take it out, you automatically shift your gait to avoid pressing on it. Pretty soon your calf begins to hurt, then your thigh, then your hip. What's happening here? To try to walk normally, you're compensating by using other muscles in ways they are not normally used, making them strained and painful.

You've been using your hands automatically since infancy. When pain from overuse disrupts the customary pattern of function, it's normal to shift to other muscles to "get the job done." But if this abnormal pattern is allowed to persist, muscle pain and soft tissue strain follow. Proper ergonomics—attention to posture, positions of tools and objects of repetitive use, workplace dynamics, matching jobs with an employee's physical size, strength, and dexterity, and a host of other factors—are amazingly helpful in intervention and healing.

Our hands are what allows us to function as human beings. With them we eat, drink, work, touch our loved ones, manipulate our environment. Loss of normal hand function is upsetting and creates a classic grief reaction, much like losing a close relative. When upper extremity pain makes work difficult or impossible, and people so affected *must* work to earn money to survive, the conflict is particularly acute. On the wall next to my exam table I keep a cartoon depicting a surgeon with his gloved hand outstretched, beckoning for an instrument. The caption reads "Magic Wand." While hand surgeons can effectively eliminate CTS numbness, nighttime awakening, and loss of strength and dexterity with an outpatient operation, the pain problem is more difficult to treat, and it is indeed magical thinking to expect a quick solution. What makes it so frustrating is the degree of work modification required. Patients and their physicians, employers, therapists, and rehabilitation specialists must all work together to ensure a satisfactory outcome.

19.

Rheumatoid Disease

The name *rheumatoid disease* is indeed appropriate. Stress the word *disease*—arthritis is only a part of this malady. Virtually every organ system is affected, separately named and often treated by different physicians specializing in whatever body part is targeted. Inflammation of the tear glands or salivary glands produces dryness of the eyes and mouth (sicca, or Sjogren's syndrome). Anemia may occur. Inflammation of the membranes around the heart causes pericarditis, and you may see a cardiologist. One in five patients develops painless but often large nodules in the skin of the hands and elbows or other areas, even internally. These rheumatoid nodules contain a core of cellular debris within a matrix of connective tissue and inflammatory cells. They are usually pain-free, but annoying and unsightly. There may be inflammation of the blood vessels—vasculitis, sometimes producing leg ulcers—or the lungs (pleuritis). Commonly, inflammation affects the lining, lubricating joint and tendon tissue—the synovium—producing a progressive inflammation, a synovitis, leading to erosion and destruction of joint-stabilizing ligaments, leading to joint deformities, tendon ruptures, and a sequential cascade of musculoskeletal abnormalities. Because the rheumatoid process is usually symmetrical, starting in the small joints of the hands and feet, it often creates pain and limitations and interferes with function.

The net effect is a potpourri of these specific as well as other troublesome generalized signs and symptoms: muscle and joint aches, low-grade fever, morning stiffness, fatigue, sleep difficulties, depression, poor appetite. Mercifully, periods of disease activity, called flares, are interspersed with unpredictable intervals of quiescence, or remissions. Either may be long lasting or short, with or without medication.

Rheumatoid disease is distressingly common. It affects two and a half million people in the United States alone, and about 1 percent of the population worldwide, with women three times more commonly afflicted than men. It can occur at any age, with a peak incidence in the twenty- to fifty-year-old group. Overall, 10 percent of those who show initial symptoms, the lucky ones, don't progress. In the group below age sixteen who develop so-called juvenile rheumatoid arthritis, more than 50 percent show no progression with time. That's the good news. If progression does occur, it's likely to last a lifetime.

Rheumatoid disease is not genetically inherited, although genetics certainly plays a role. Siblings in families in which the disease is present have a four to six times higher risk of contracting the illness. As is probably true with many illnesses there is, presumably, a genetic predisposition, but the disease doesn't manifest itself unless there is also an environmental trigger. It's not known precisely what the trigger is, but it's currently felt to most likely be an infectious agent— a virus or a bacterium (unaltered or altered, Leviton's "stealth pathogen"). If one monozygotic twin has rheumatoid disease there is only a 15 to 20 percent chance the other twin will have it.

What does seem clear is that the immune system is affected in such a way that the afflicted individual fails to recognize specific body tissues as "self" and manufactures antibodies that immunologically attack these tissues. This defines a diagnosis of autoimmune disease.

Unfortunately, the chief target of the autoimmune attack is the synovium, the filmy tissue present throughout the body, which, as noted, lines and lubricates the joints, and that surrounds and lubricates the tendons where they pass through tight pulleylike areas and must turn corners to be functionally effective. Synovium normally manufactures a high-grade oil with superb lubricating characteristics. When affected by the rheumatoid onslaught, synovium loses its lubricating prowess and becomes boggy, swollen, and thickened and weeps a fluid rich in digestive enzymes and inflammatory, tissue-destructive chemicals.

The Latin root words taught to all medical students to describe inflammation's characteristics are: rubor (redness), calor (heat), dolor (pain), and tumor (swelling). As if on cue, affected joints become red, warm, painful, and swollen. They tend to stay this way unless some intervention—drugs or surgery—aborts the process. The Latin suffix *itis* means inflammation, hence appendicitis, laryngitis, arthritis, vasculitis, pericarditis, synovitis, etc.

Moreover, the destructive chemicals manufactured as part of the inflammatory process weaken or actually dissolve adjacent bone or connective tissues, like ligaments or cartilage (which is marginally nourished normally and is a sitting duck for this sort of badness). Bones become pitted and eroded. Cartilage disappears and, as it does so, the joints shorten as their lining surfaces vanish. Normally tight supporting and stabilizing ligaments on the sides of the joints become abnormally loose and incompetent, allowing intact tendons to pull the joints into odd positions. Voilà, deformity. The definition of a deformity is an imbalance of forces acting around a mobile joint. The tendency of the fingers to deviate toward the little finger side—so-called ulnar drift—is the best-known example of this kind of rheumatoid deformity. There are others.

Because this is an immune disease, it is not surprising to see specific antibodies appear in the blood as markers of the ongoing battle. The "rheumatoid factor" appears in 80 percent of patients. A useful blood test of disease activity is the sedimentation rate, or "sed rate." This simple test measures the propensity for red blood cells to clump and settle in a test tube as a measure of active inflammation and is elevated during a flare. These tests are the diagnostic backbone used by all physicians. Rounding out the battery of helpful studies are X rays of affected joints and analysis of fluid obtained from joints. The procedure, called arthrocentesis (joint tapping), involves placing a sterile needle in a swollen joint and drawing off fluid. This may be helpful both for allowing direct analysis of the fluid as well as for treatment, to relieve painful swelling and to allow instillation of cortisone (the quintessential anti-inflammatory drug), especially in large joints.

Rheumatologists who treat rheumatoid disease have at their disposal a battery of drugs formulated by the pharmaceutical industry to combat this ailment. When I was a medical student assigned to a rehab service, I helped care for an elderly woman, stone deaf from taking twenty-four nonenteric-coated aspirin a day for her rheumatoid arthritis. Fortunately, she did not have a sensitive stomach and had no problems with GI bleeding. Aspirin is the classic anti-inflammatory drug and little else was available at the time. The deafness is reversible and goes away if the aspirin dose is reduced; hers did.

Drug therapy has made great strides in the past twenty years. We have a large variety of NSAIDs (nonsteroidal—i.e., noncortisone—anti-inflammatory drugs). Some actually have no irritative effect on the GI tract. NSAIDs are well tolerated, with few side effects, and constitute the first line of defense in rheumatoid disease.

The second line of defense, a really powerful one, are the so-called DMARDs (disease-modifying anti-rheumatoid drugs) like corticosteroids (cortisone), gold (found many years ago to slow the progression of rheumatoid disease, although chronic absorption of gold salts by mouth turns the skin a metallic silvery color), methotrexate, sulfasalizine, d-penicillamine, hydroxychloroquine (Plaquinyl or other antimalarials), Imuran, Cytoxan and, now, Etanercept (Enbrel, FDA approved since November 1998).

With time, several things have become clear. Rheumatoid disease is not benign. It provokes nine million physician visits annually in the United States and a quarter million hospitalizations, at a staggering cost of over $60 billion. Approximately 50 percent of patients are work disabled within ten years of diagnosis, at an annual cost in lost wages in the United States alone of nearly $7 billion (1986 dollars).

The first-line NSAIDs, while generally well tolerated, are not without their own toxicities. The second-line DMARDs are not as toxic as previously thought. It's a hard lesson but, once joint erosion and functional loss have occurred, they may not be recovered or reversed. So there is an increasing tendency to use DMARDs early in the disease, along with NSAIDs, to try to forestall the evolution of irreversible problems. Moreover, there is no clear consensus—yet—over which of the DMARDs to use. All have recognized toxic potential, but all can be used with relative safety for about three years, after which there is a tendency to switch or rotate them.

On the horizon are new approaches. These include: monoclonal antibodies to combat certain white blood cells active in the inflammatory cascade; and monoclonal antibodies against tumor necrosis factor (TNF alpha) felt to be especially active and destructive in rheumatoid inflammation (Enbrel binds with TNF making it inactive, reducing joint inflammation). New pharmaceuticals, both NSAIDs and DMARDs, are constantly under development. It is hoped these will prove to be effective. Because, at least for the moment, there is no cure.

However, for someone like Mona, Connie's Christian Science sister, in chapter 8, "them's fightin' words." A firm believer in the power of personal conviction, in her confidence that any ailment will cease to be a threat once assailed with an overwhelming barrage of optimism, she refused to willingly yield control of her sister's medical care. Perhaps not surprisingly, Mona's holistic approach—shunning surgical or drug-based treatment protocols in favor of dietary or other systems—has, if one searches for it, some data to support it. The support comes from unconventional though widely circulated sources.

Health data abounds. As reported by the Optimal Wellness Center, a study from the National Public Health Institute in Helsinki, Finland, followed nineteen thousand healthy men and women for fifteen years. During that interval, 126 people developed rheumatoid disease, nearly three-quarters of them positive for rheumatoid factor. Consumption of more than four cups of boiled coffee daily was associated with an increased risk of contracting the illness. Boiled coffee contains many chemicals removed by filtration; it is acknowledged that the routine practice of coffee filtration may lower the risk.

Cigarette smoking also seems to be associated with increased rheumatoid risk. In a study published in the *Annals of the Rheumatic Diseases*, Dr. Kenneth Saag, Assistant Professor of Internal Medicine at the University of Iowa, found heavy cigarette smoking worsened the rheumatoid disease of 336 patients studied and was associated with a higher level of rheumatoid factor and a greater degree of bone erosion.

So should Connie give up her pack and a half a day habit (not so easy to do) and throw away her "bottomless" coffeepot solely to improve her rheumatoid disease? It's easy to say yes because of the health benefits accrued, whether the rheumatoid disease improves or not. But such a vigilante approach fails to allow a study, well designed to document these benefits, to stand on its own.

More health risks for rheumatoid sufferers. Food allergies are felt by some to play a significant role in worsening the disease. Cow's milk and wheat products are big offenders, as are corn and soy products, nightshade vegetables (tomatoes, particularly if green, and eggplant), sweet or fatty foods, and, especially, polyunsaturated oils.

Rheumatoid patients are seemingly beset with numerous digestive disorders. The range is prodigious: medication-induced intestinal wall malabsorption; low gastric acid levels; unhealthy intestinal flora leading to "sludged blood"; calcium leaching out of bones, joints, and teeth; inadequate minerals and trace elements; dangerous microbes; and poor elimination. "Bowel cleansing" is often recommended. This is accomplished with laxative foods and ingested Epsom salts, ground linseed, aloe, or other agents sufficiently active to produce two to three bowel movements daily. Fatty and fat-soluble wastes and toxins are thereby washed out of the GI tract.

The list of environmental toxins and hazards as rheumatoid risk factors doesn't end here. Add to these: solvent fumes, mothballs, gasoline fumes, makeup, nylon, detergents, dishwater, and synthetic or agricultural chemicals.

Conversely, certain food substances are thought to be beneficial. More than thirteen Australian studies have trumpeted the worthwhile aspects of fish oils in ameliorating morning stiffness and disease severity after fifteen weeks, as reported by patients and their physicians. Chiefly targeted for reduction were omega-6 oils felt to be inflammatory through their eicosanoid metabolites. With a dietary shift toward omega-3 and away from omega-6 oils, benefits for rheumatoid patients mounted. Reinforcing this view have been studies from Spain, Belgium, Boston, Albany, New York, and Florida, all concluding that, after twelve weeks of supplementation, fish oils, chiefly omega-3, reduce rheumatoid symptoms. If one then adds to this nutritional regimen herbal remedies, essential oil massages, carefully conducted exercises, and guided visual imagery, there may be a remarkable improvement in well-being.

Even respected and widely read magazines like *Arthritis Today* have praised the advantages of "integrative medicine," a holistic approach for treating the whole person, as if conventional allopathic medicine never did. Alternative health care is growing in popularity because it is cheaper and more accessible than regular medical care and, frankly, generally more nurturing, a factor not to be dismissed lightly.

Philosophically, integrative medicine's ideology is beyond reproach: focus not on the disease, but the person who has the disease. Recognize the unity of mind, body, and spirit in a single individual facing a failure in normal bodily functioning. In treating the problem, bring to bear not only the best and latest offered by modern technology but a deeply insightful assessment of the person's needs, wishes, and psychological makeup and functioning. Infuse into the mix a measure of caring, hope, and education.

What is arguable is the panoply of well-meaning but inadequately tested or simply untested methodologies used to implement the philosophy. Nutritionist Walter Last has described a few, like kerosene plasters and vomiting therapy. Color therapy requires shining an intense blue light at close range onto a painful or inflamed area. Blistering or counterirritation therapy involves exposing an inflamed joint to sufficient heat or irritant to produce a blister. The theory is that this draws the toxins and congesting enzymes responsible for the malady to the surface. In the alternative, one can draw the toxins to the surface with ants, nettles, or bee stings, or apply a variety of poultices containing peppers or blistering agents. Either a blister or pus-filled pustules should appear as a sign the treatment is working. It is recommended to cover the area with a fresh cabbage leaf, or drain off blister fluid with a needle. Such folk remedies have been used for centuries to treat arthritic manifestations, supposedly with success, more quickly and reliably than with other methods.

I never learned about any of them in my medical training.

In a speech to the Harvard Alumni Association in May 2000, Saul Green, Ph.D., cautioned all ill people to question their health care providers and ask, "Why should I believe you?" He pointed out facts that many of us know but often choose to ignore. Example: dietary supplements are unregulated, and no FDA regulation mandates accurate or honest accounting of ingredients claimed to be present in supplements; often they are not even included, or not in the concentration or amount proclaimed on the label. The "scientific" bases for acupuncture, for naturopathy, or many other systems lack true scientific validity. Dr. Green also seriously questioned the claim that alternative medical therapies influence "healing." What, he asked, do we mean by "healing"? The impression gleaned from Green's argument is, if a methodology lacks scientific reproducibility and proper testing, scrutiny, and skepticism, it lacks merit.

But there is a mystery here, and those who can't quite abide it don't sleep well at night. Deepak Chopra was getting at it when he wrote in *Quantum Healing* of the portion of our everyday expectations that dips "below the line." It is the irrational, the unfathomable, the ineffable. We're uncomfortable with it because we don't quite know what to do with it, so we're at a loss. Does this negate its presence or its value? We can't quantify its boundaries, its limits, if there are any. Should we give up, just because it's beyond our current knowledge, or even our current ability to know?

A serious scientific group, with a membership of over 1,500, is ISSSEEM, the International Society for the Study of Subtle Energies and Energy Medicine. ISSSEEM is an interdisciplinary organization dedicated to improving human health and welfare through education, training, research, and practice in the emerging discipline of energy medicine. Energy medicine deals with the notion that many of the currently unmeasurable phenomena associated with health and healing involve energies (such as, perhaps, the HEF—human energy field popularized by Dolores Krieger) that will in the future be routinely measured. This society features lectures and workshops with prominent M.D. and Ph.D. scientists actively involved in this work. These include: Norman Shealy, M.D., Ph.D., a neurosurgeon who now is well known for his work in pain management; Robert Becker, M.D., former Chief of Orthopedics at the University of Syracuse Medical Center, pioneer in research into limb regeneration in mammals, and currently studying the effects of exposure to high-voltage electric lines; founding president Elmer Green, M.D., biofeedback pioneer; and Larry Dossey, M.D., author of *Healing Words*. ISSSEEM's well-published leaders are no slouches. It's just that we're perilously close to the unknown in exploring the subject of energy medicine.

But how can such energies be quantified or measured? How does one design a proper experiment, asking answerable experimental questions, with appropriate controls and manipulation of single variables?

Does prayer work for rheumatoid disease? And if so, by what criteria do we measure its effectiveness?

In recent years, bold investigators have risen to the challenge of trying to answer these questions, with mixed success. Cardiologist Randolph Byrd exposed 393 patients in the coronary care unit of San Francisco General Hospital to intercessory prayer, that is, prayer from a distance. The investigators claimed the study was double blind—neither the patients nor the medical staff caring for them knew who was the object of prayer. This 1988 study, published in the *Southern Medical Journal*, showed strong benefits from this form of prayer. Internist Larry Dossey, in *Healing Words*, criticizes the absence of life and death results in this study, but praises it as the "seminal event" stimulating his own endeavors to document prayer's healing powers. Byrd's study certainly stimulated interest in approaching these questions seriously.

Later, Dossey surveyed over 130 controlled experiments, showing that intercessory prayer helped rye grass grow taller, helped yeast to resist the toxic effects of cyanide, and helped test tube bacteria to grow faster. Dossey was enamored of these experiments because they did not involve humans, with all their messy, hard-to-control variables; they could be replicated an infinite number of times; and they worked anywhere on earth they were done.

Georgetown University Medical School professor Dale Matthews, M.D., conducted a 1996 study in Clearwater, Florida, of rheumatoid patients. All patients were first subjected to a hands-on "faith healing" session. They were then split into groups, some of whom (the test group) received intercessory prayers from a designated priest and some of whom (the control group) did not. The study was subjected to media coverage and there is some confusion about numbers and specific protocols, violation of control criteria, and evaluation of results. There were suggestions that improvement from prayer was dramatic and sustained in many cases, but firm conclusions could not be drawn.

"Prayer works," says Dr. Matthews, author of *The Faith Factor: Proof of the Healing Power of Prayer*. Thirty U.S. medical schools as of March 2001 were offering courses in faith and medicine. Jeffrey Levin, Ph.D., associate professor of family and community medicine at Eastern Virginia Medical School in Norfolk, has said "all types of prayer appear to work." Both he and Dr. Dossey agree empathy is prayer's key element. "There has to be caring," says Dossey. "The desire for recovery has to be genuine, authentic and deeply felt. It has to come from a feeling of love and compassion." Says Matthews, "Ministry is a complex intervention that combines touching, caring, listening and praying. It's the whole package."

Not all are convinced. Internist Joel Taurog, M.D., professor and interim chief of the rheumatic disease division at University of Texas Southwestern Medical Center, has pointed to numerous design flaws in Matthews's study: improper controls, lack of randomization, and failure to account for the placebo effect.

But the patients who were prayed for got better and stayed better, an observation that encouraged Matthews. The fact that serum markers for inflammation did not change, even though most patients experienced improvement in swelling and tenderness, suggested to him that improvements were mediated by a different mechanism than the inflammatory pathway.

"I wouldn't dismiss the results as meaningless," says Taurog. "It's an interesting study that has some flaws, but still merits further investigation."

So where are we? It seems certain that with interest stimulated, research in this area will continue, and, one hopes, will be of good quality with flawless design standards. Dossey thought the beneficial effects of prayer did not require an intermediary but came about as the product of one person's thought waves interacting with another's body. The measurement, or even proof of existence, of such psychic powers is beyond the scope of current science.

But why not? "Healing" can certainly be experienced subjectively as well as observed and studied objectively. If we are careful with definitions and descriptors; talk about "improvement," however we choose to classify it, rather than "cure"; and include the entire spectrum of subjective criteria used for evaluation, "healing" goes far beyond Dr. Green's narrow definition and takes on a more humanistic—and truthful—cast.

It's depressing to have rheumatoid disease and it's depressing to be limited in what you can do. The majority of patients are women and, as reported in the May 2001 issue of the *Annals of Behavioral Medicine*, as many as 25 percent of women will suffer from depression during their lifetime, making this the most common affective disorder. With depression goes fatigue, and debilitating fatigue is 10 percent more common with rheumatoids than in other illnesses. This is familiar baggage for rheumatoids to carry.

The temptation is strong to take antidepressant or mood-elevating drugs and put up with the side effects. But research has compared the effectiveness of these drugs with so-called cognitive therapy—in effect, talking yourself out of it. While this may at first seem like a juvenile approach, what is prayer after all, and how is this different from prayer? People who truly believe they can feel good seem to better remember what feeling good was like and, if motivated, can bring back those memories. Deep faith is calming and encourages discipline if lifestyle changes are needed.

It would be adolescent folly to presume the answers are easily arrived at, but they are there, and when they are discovered, they will be first denied or doubted, then widely disseminated and used, then trivialized. It has happened repeatedly before, and it is human nature to do so.

20.

Thumb Aplasia: Missing Thumbs

Both human and some subhuman primate infants demonstrate in utero a consistent activity. They show a repeated pattern of hand-mouth associated motions. That is, the hand is frequently— more than fifty times an hour, based on ultrasound studies—brought toward or against the face and, if it contacts the face in proximity to the mouth, the mouth makes sucking gestures or opens. Extensive research has documented the idea that many nonspecific reflexes—like the rooting reflex (the reflex of opening the mouth to nurse when the cheek near the mouth is stroked)—are not responsible for this behavior. It appears this innate coordination is a product of gustatory, or feeding behavior that occurs only when the infant desires food. Such a sequence in turn depends on a sequential development of complex neural pathways, now well defined, in the neonate's brain. Sucking on the hand seems to be a characteristic of primate maturation. It continues after birth, occurring typically in hungry neonates prior to, but not after, feeding.

But what if there is a catastrophe and the hand fails to develop? What happens to this dependable, specific reflexive hand-to-mouth movement?

The answer is astonishing. In the absence of a real hand, the brain may create a phantom of the missing hand, and attempt a "business as usual" behavior pattern. What is more, available evidence suggests both sides of the circuit are represented. That is, even if the hand is absent, the mouth responds to a "phantom hand" as if a real one were present. Hunger triggers activation of a well-defined nerve pathway sequence, even if one critical anatomic portion of the sequence is simply not there.

The existence of a so-called aplastic phantom triggers a host of questions. One of the most fundamental is whether such evidence confirms the presence of a body image as opposed to a "body schema." To neurobiologists, the distinction is crucial. Body image refers to a passive structure, a system more like Penfield's homunculus, more like a static and unmoving photographic representation hung on a wall.

In contrast, a body schema is more active and relates to one's capacity to move, to do something. It has to do with motor capacities, habits that enable movement or maintenance of posture.

Important in sorting this out is the phenomenon of "forgetting." Limb amputations almost invariably lead to the appearance of a phantom limb. The phantom is often so vivid, so real, that the affected person may "forget" the real limb is gone and may, unsuccessfully, attempt to use the phantom for functional tasks. This kind of functional hallucination probably represents the body's attempt to maintain a well-rehearsed status quo and, in the attempt, it simply "forgets" about the loss.

Significantly, no examples of forgetting have been seen with aplastic phantoms. The commonly drawn conclusion from this observation is that the creation of a phantom is more functionally accurate, more reality based after traumatic loss, especially if the individual has had experience using the extremity in daily use.

Despite this, hand-mouth coordination movements are present even when no hand exists, for any reason—trauma or birth absence. The intensity may decay with time, suggesting that some degree of touch balance feedback is required to maintain a strong reflex. If the hand portion is not there, the intensity of the reflex deteriorates because one loop of the circuit is not constantly activated. Nevertheless, even with hand aplasia, the brain's enabling neural circuits are preserved and are present, even if not used to their normal extent. These observations reinforce conclusions drawn by Gallagher: the hand-mouth coordination seen in infants supports the notions of *both* body image and body schema, and *both* are probably operational in humans.

Implications that flow from these observations and supporting data suggest that the brain's internal representation of the hand is genetically determined, and that the neural matrix responsible for coordinated and complex hand activity is widely distributed through the brain, not restricted to only one limited area. Because of repetitive feedback stimulation of familiar motor sequences, traumatic loss of a hand (as with amputation) results in extensive, wide-ranging reorganization within the brain to keep things on an even keel. Such "plasticity" occurs promptly in developing embryonic brains. New evidence, amassing almost daily, suggests the same phenomenon occurs in adult human brains, albeit much more slowly.

Plasticity refers to the brain's lifelong ability to adapt and change. The brain is dynamic, flexible. New scans of brain activity, such as PET scans, which can show metabolic or chemical activity, reveal furious turnover in response to new situations.

A child learning to play the violin demonstrates markedly increased activity and enlargement in the part of the brain controlling the fingers of the left hand, in which fretting the strings assumes prime importance. The left hand's activity requires strength and dexterity. The brain obliges by doing its share, enlarging the portion of the motor cortex operating the involved fingers. On the opposite side of the brain, which operates the bow hand and in which such modifications are not required, these changes do not occur.

In blind people who learn to read braille, the relatively unused portion of the visual cortex enlarges and is taken over by neurons operating the reading fingertips and assisting with the sense of touch.

Japanese research focused on a deaf man watching videotapes of people using sign language. Brain scans showed a portion of the cortex normally used for hearing was "recruited" to help him decipher sign language.

It seems, therefore, that there is much more for the brain to do—neuronal branching and formation of new synapses—after traumatic loss of a hand than with aplasia. If generation of a phantom is a compensatory brain activity, one would expect phantoms to be less common when limbs are missing due to aplasia. This fits well with behavioral data

Cases like Ricky's—he was born without thumbs—have not been studied as exhaustively or with the same finesse as I have described. Based on such sophisticated, esoteric information, and extrapolating to his situation, it makes sense on a neurological basis that the finger-to-thumb transfer should work as well as it does. Since the brain expects a digit in the thumb's position to behave and work like a thumb, hooking up a digit with the proper connections in the thumb slot becomes, in a way, a self-fulfilling prophecy. The same can be said of the technique pioneered and championed by California surgeon Harry Buncke, the big toe-to-thumb transfer. Moving a big toe to the hand obviously had to wait until microvascular surgery became consistently successful enough to make such treatment a practical reality. Now that it has, the purists in the crowd may argue a great toe on the hand still looks like a toe, not a thumb. But it can sure work well and overcome the many functional limitations inherent in the absence of the thumb.

21.

Replantation; Leeches; Prostheses

By general law, life and limb must be preserved, and often, a limb must be amputated to save a life, but life is never wisely given to save a limb.

—Abraham Lincoln

I am always doing that which I cannot do, in order that I may learn how to do it.   —Pablo Picasso

Once considered pure science fiction, technological advances in modern surgery have made replantation of hand and fingers commonplace. Of course, there are limitations and contraindications. The secret of success is knowing what these are and obeying the rules they impose.

Even the word *success* carries elusive connotations, in large measure because of unrealistic expectations on the part of the patient, family, and friends. What does success mean in replantation?

First, and perhaps foremost, the replanted tissue must survive.

Survival has become all the rage in popular culture, and the general concept has spawned a successful TV show. But applied to the technical details involved in making amputated human tissue (dead without big-time help) survive, the word carries more subtleties than most people realize.

It is a natural, normal desire to be whole, to be intact. We all carry a "body image" in which our parts are intact and in place. So when trauma, surgery, or other life events remove a part, we long for it and want it back. In our society, the technological stewards of biologic preservation are the physicians. So it has fallen to them to figure out how to pull it off. It's been a long and rocky road.

Even in ancient times, the desire was strong to keep the body in its pristine state. But if a finger, a hand, or an arm was removed, usually traumatically, through accident or deliberately as punishment, there was simply no way to put it back. Surgery, as we know it, did not even exist then, and anesthesia wasn't discovered until 1847. So a variety of herbal and natural plant products was used for curative intervention: nepentag, opium, hemp, mandrake, henbane, hemlock, and, of course, alcohol. Asperic acid from tree bark was used as an analgesic to relieve pain. Antiseptics like smoke, honey, and wine were applied to wounds, but the germ theory was unknown and medical luminaries like Ambrose Pare referred to "laudable pus" in the 1500s and poured boiling oil on wounds, thinking it was a good idea (it wasn't).

William Balfour is credited with performing the first successful fingertip reattachment in 1814. Thomas Hunter successfully reattached a thumb tip one year later. In each case, the reattached tissue survived as what is known as a *free composite graft*, winning a race against time, living because new blood vessels from the digit grew in to nourish the tissue before it could die from lack of nutrients and oxygen. This is a far cry from true tissue replantation, in which larger body parts survive because their (usually tiny) blood vessels are repaired.

Such vascular (blood vessel) repairs could be consistently achieved only when technology caught up with desire. The work has formed a continuum of pioneering accomplishment. William Halstead and Alexis Carrel performed replantation experiments with dog limbs in the 1880s. This work paved the way for kidney transplantation, which hinged on vascular repair. For it, Carrel was awarded a Nobel Prize in 1912.

Organ transplantation, now considered so routine it barely merits a ho-hum from the general public, could not exist without a well-worked-out schema for organ preservation: cold storage (cold slows down the inevitable march toward cell death), consistently successful vascular repair, and sophisticated anti-rejection drugs. Everything is set up in a hospital operating theater to ensure maximum efficiency. Especially in adults, the main arteries and veins to and from organs like kidneys, livers, and hearts are pretty good-sized. For a technically competent, well-trained, and practiced surgeon such repairs can be achieved consistently. Even sewing in the smaller vein grafts used in coronary bypass operations has become routine, belying the surgical skill required. As the admonition goes, don't try this at home.

Replantating severed digits, though, is something else. First of all, conditions are generally far from ideal. The amputation is almost always traumatic, often messy, with a great deal of associated tissue mangling (as in Kathy's case). Amputated parts may not be recovered promptly, often are roughly handled, washed in tap water (which is hypotonic—lacking the normal balance of salts and chemicals found in blood and serum—and kills cells on contact), and are rarely kept cold in a non-injurious way—placed in a plastic bag or clean jar inside an ice-filled container without allowing the amputated part to rest in contact with the ice. Most commonly, the location of the injurious event is some distance from a medical facility and, by the time all appropriate resources have been mobilized, one is really pushing the clock.

So just getting to the stage of vascular repair is difficult. Remember, I said success is elusive. Even if it's just one digit, bones need to be properly aligned, with correct rotation, and fixated. Tendons need to be sutured together to provide the opportunity for movement. And nerves need to be repaired to allow for the possible recovery of feeling, all before the skin can be closed.

All tissues heal together with scar—nature's glue—and the effect is quite literally like pouring glue into the wound. Everything sticks to everything else, and attaining differential gliding, so flexor and extensor tendons move separately from each other and the finger will bend and straighten, involves a lot of therapy and work. If the nerves regenerate, recovery of feeling takes months to years and is never normal.

Thumbs are unique in their function, and are considered special enough to warrant attempted replantation. But a single finger? Most centers capable of doing it will not replant a single amputated digit. Experience has shown it is more important to preserve normal use of the remaining fingers. In a community of digits, one digit can, and should, be thrown overboard to ensure proper functioning of the group.

And yet, replantation has captured the imagination of the public, partly because it's a challenge, partly because finger amputations are so painfully common. Despite the strong motivation to re-create wholeness, reliable repair of finger blood vessels, only one or two millimeters in diameter, had to wait until technological advances could catch up.

In 1921, Nylen introduced the operating microscope for ear surgery. This was more of a research tool, not widely or commercially available for all applications. In 1953 Zeiss Corporation mass produced a microscope usable for replantation work, but it had numerous limitations. It was not until 1960 that Jacobson and Suarez introduced an instrument similar to the one used today, with better magnification and focus. Even so, repair of transected digital vessels carried only a 50 percent success rate.

Progress exploded rapidly. In 1962, Ron Malt, a general surgery resident at Massachusetts General Hospital, was dispatched in an ambulance and brought in a twelve-year-old boy with his severed right arm, cut off in a train accident. Malt, with no preconceived ideas about his historic venture, persuaded a multispecialty team of physicians to participate in a "wonderful experiment" and reattach the limb, ushering in the era of microvascular replantation. Malt, in 1995, wrote to New Zealand hand surgeon Martin Entin: "I hate to think what might have happened if we had bungled."

What followed was an international cascade of surgical efforts and communications. In 1965, S. Tamai replanted a human thumb. Three years later John Cobbett completed a big toe-to-thumb transfer. But great credit is due American surgeon Harry Buncke, who devoted years of research and effort to making true microvascular surgery an everyday reality, completing his own first big toe-to-thumb transfer in 1972 and making this procedure practical.

China, in the throes of perpetual modernization of its industries, had primitive machines with inadequate safety devices. Consequently, and not surprisingly, China accumulated a near universal monopoly of upper-extremity amputations in the 1960s. Spurred on by Chairman Mao's priority of "salvaging our workers' hands," the Chinese startled all advanced nations with their huge volume and fine results following replantation, using older tools and magnifying loupes instead of microscopes.

Such excellent results (50 to 60 percent survival achieved by the Chinese) were considered first-rate at that time. Survival, however, increased to over 80 percent as a result of the teaching of Chairman Mao, who counseled his followers in his *Little Red Book*: "In dealing with forces of unequal strength, one has to strengthen the weak to overcome the strong." Shanghai microvascular surgeon Ch'en Chung Wei applied this principle to his finger replantations, which often darkened and died, frustratingly, after initially looking pink and healthy. It was then the custom to repair one artery and only one vein. Realizing venous drainage was the "weak force" that needed to be strengthened, Ch'en began a policy of repairing two to three veins to allow blood to drain out of the finger and prevent congestion. Survival of replantations improved dramatically once this modification was implemented.

In a similar vein (no pun intended), an ancient technique has found resurgence supporting the microvascular craze: medicinal leeching. Draining blood to cure disease is an ancient custom dating back to the Stone Age. Use of the leech *Hirudo medicinalis* to draw off blood probably originated in China and India. A painting in an Egyptian tomb built in 1500 B.C. shows leeches being used medically. Chinese, Sanskrit, and Arabic writings two thousand years old describe it. The great physician Galen promoted bloodletting in his humoral concept of disease, maintaining, in the second century, that one could restore the balance of the four humors—blood, phlegm, black bile, and yellow bile—by drawing off some blood.

This was common practice for hundreds of years, through the Middle Ages and Renaissance, but with one glaring problem. The knives, bones, and sharp sticks used for curative bloodletting often let off more blood than was deemed necessary for cure. George Washington died a day after overexuberant bloodletting was used to treat his laryngitis. A gentler—and more consistent—method was needed. Thus the popularity of leeches grew, not because they were cuddly— most people think a four-inch-long leech is repulsive—but because they were less painful and reliably withdrew only about two-thirds of a tablespoon (10 cc) of blood.

Leeches had been in abundant supply in Europe but, to support agriculture, drainage of wetlands where they thrived combined with heavy medicinal use produced major shortages. In 1833, a peak year, France was forced to import 41 million leeches to keep up with demand. In America, leeches had been credited with saving many lives during the yellow fever epidemic of 1879. When shortages grew, despite waning demand, medical suppliers were forced to use the smaller local species, which drew off less blood.

By the late nineteenth century, demand for leeches decreased as improvements in medicine made bloodletting less desirable as a remedy. Medical use of leeches might have faded into obscurity but for the discoveries of John Haycraft in 1884. Haycraft found that an animal's blood in a leech intestine doesn't clot, and a wound made by a leech bite bleeds longer than normal. Such seemingly innocuous observations were not overlooked and led to a closer investigation of leech saliva. By the late 1950s, a powerful anticoagulant, hirudin, was isolated from leech salivary glands, ensuring a place for the humble leech in the modern medical era.

In fact, careful study has revealed the leech to be a storehouse of unique chemicals. Hirudin has been purified and cloned and is now commercially available as a recombinant protein, approved by the FDA as lepirudin. Numerous other leech salivary proteins and enzymes have been shown to have effects in the blood-clotting cascade, the known sequence of chemical and cellular steps required to convert liquid blood to a clot. These not only prevent coagulation but dissolve clots, inhibit platelet aggregation, prolong bleeding time, and have many other actions. All are under investigation.

Because of these properties, leeches have now found an established place in the replantation armamentarium. If replanted tissue becomes engorged with blood, usually because of inadequate venous drainage (as Dr. Ch'en noted), leeches drain the tissue and prevent clot formation until normal venous drainage can be established. Leeches can improve blood flow and microcirculatory performance. The anticoagulant effect is purely local, confined to the tissues being treated, without affecting the whole person. It takes one leech under an hour to consume two teaspoons of blood. The leech then drops off and may not need to feed for nearly a year.

Leeches aside, establishment of replantation of severed parts and organs as routine procedures paved the way for the next technical leap: transplantation of a limb from one human being to another. Of course, all donated organs like kidneys, livers, and hearts are transplants from one human being to another. Once the technical obstacles in actually doing it have been overcome, the biggest problem in having the transplanted organ work is immunologic rejection. Our immune systems, unfailingly protective of us, recognize the foreign organ or any foreign tissue as "not self" and mount a fierce internal biochemical battle to cast off the invader, a process known as rejection.

Human tissues contain antigens, both internally and on their surface. The most important in establishing compatibility between donor and recipient are the so-called HLA antigens, because they were first described in human white blood cells (leukocytes). There are three classes of HLA antigens, located on different cell types, but the first two are most active in transplant rejection.

Rejection affects all tissue transplanted across immunologic barriers (least of all from one identical twin to another; in this one situation, there is immune tolerance). Tissue typing to establish biocompatibility for organ donation and transplantation has become ultrasophisticated, as have the new drugs used to suppress the immune system and prevent rejection. This is a dangerous game, since we need a functioning immune system to ward off life-threatening infections or other microbial invaders.

It is now possible to transplant an internal organ from one human to another and have it survive and function. Organs that are relatively docile immunologically are bone, corneas, heart valves, and preserved tendons, probably because all have relatively few cells and express antigens weakly. Other relatively homogeneous organs are more difficult, but manageable. With them, for the most part, the body is dealing with only one type of tissue—be it liver, kidney, or whatever.

It's much more difficult to accomplish the same ends with an arm/hand transplant, for several reasons. The first is that such transplants contain multiple, diverse types of tissue—nerve, blood vessels, fat, tendon, muscle, and, especially, skin. Skin is particularly active immunologically and is violently rejected when grafted across immunologic barriers, as from one person to another.

A second reason is the functional expectations. A transplanted kidney simply has to make urine and, in doing so, clears toxins from the body. A heart has to beat and propel blood. How simple. In contrast, a transplanted arm is expected to have feeling (never normal, and nerve regeneration takes months to years); have movement (never with perfect control and, only with great good fortune, specific individual muscle control); and survive, free of rejection, which, if it occurs, may significantly impede nerve and tendon function.

Despite all the obstacles, arm transplantation is on the cutting edge of medical technology and, from a human interest point of view, makes good press. The world's first human arm (forearm-level) transplant was carried out in France in September 1998. The recipient was a forty-eight-year-old Australian, Clint Hallam. While initial results looked optimistic, Hallam traveled quite a bit on his own, without medical supervision, to the consternation of his physicians, and was cavalier about taking the new and powerful drugs given him to combat rejection. Ultimately, he lost the rejection battle and, on

February 3, 2001, after less than three years, had to undergo removal of the transplant.

In January 1999 the first hand transplant done in the United States was carried out on thirty-eight-year-old Matthew Scott in Louisville, Kentucky. Scott has been much more compliant than Hallam and so far is showing good functional recovery with no serious rejection episodes.

To date, over a dozen hand transplants, some double hands, have been carried out worldwide, including a second one in France. The Louisville team completed its second hand transplant on thirty-six-year-old Jerry Fisher in mid-February 2001, and plans ten more.

Many surgeons and biologists have expressed harsh disapproval of these hand transplantation efforts. They cite the virtual certainty the recipients will need to take potentially life-threatening anti-rejection drugs for a lifetime, and claim such treatment should be reserved only for life-threatening conditions. These claims are well based, well intended, and defy criticism. On the other hand, when asked, Matthew Scott loves his new hand, and having it has changed his life. "The hand is absolutely part of me now. This is the greatest thing that's happened to me." He is strongly motivated to keep it, at all costs, and seems willing to accept the risks.

In 1981 my wife and I visited Fontainebleau Palace outside Paris. It was a drizzly day and nearly everyone had a raincoat. In one of the painting-filled bedrooms I stood next to a slim young woman who carried a light trenchcoat draped over her left forearm from elbow to wrist. The pose was natural and I didn't give her a second glance.

As the morning wore on, I kept encountering this young woman tourist moving with the crowds from room to room. By the second or third time, I noticed her pose never changed. Even when the sun finally came out, she walked with her elbow bent, the coat folded naturally over her forearm, which she held against her body. Her hand rested in a relaxed posture, fingers slightly bent.

Curious, I revealed my suspicions to my wife (who by now wondered why I was so interested in this woman) and we surreptitiously followed her. From one ballroom to the next, her arm and hand never changed or moved. The hand was finely detailed, a perfect color match with the right, soft-looking and entirely natural.

It was years before I realized I had been admiring the creation of Jean Pillet, the world's finest hand prosthetist. Made from sculpted silicone rubber, now fashioned entirely from medical-grade silicone, superbly matched to her normal right hand in all respects, the prosthetic hand was a work of art equal to the best we saw in Fontainebleau. It passed the most meticulous inspection.

Except for one thing. It never moved.

The young woman could "fake it" easily. Her trenchcoat hid the telltale join between the prosthesis and her stump. The position of repose worked well for a casual glance. For walking around and looking acceptable, for passing muster at a wedding or social gathering, it was a winner.

For using what remained of the hand, though, this elegant, expensive prosthesis was useless. Even to make the stump work as a helper hand, sensibility—feeling—was required. In private, the prosthesis went into a drawer and the stump, covered with sensitive, nerve-containing skin, would hold down a piece of paper so the intact hand could write. And—critical for normal upper-extremity function—do it automatically, without having to look at it. When the thumb is preserved but all four fingers are gone—a distressingly common injury—an elegant prosthesis can be made to act as a fixed stump against which to squeeze objects with the thumb. But without feeling to provide feedback to know how hard you're pressing, you have to look at it. You can't use it with true automaticity.

The human desire for "wholeness" is both ancient and irresistible. Even the youngest child drawing stick figures knows there are five major human appendages—a head, two arms, and two legs. Since loss of the head is (at least, for now) incompatible with life, we don't get into discussions about prosthetic replacements for this appendage. However, loss of arms or legs is as old as human history, and the subject of endless and well chronicled scrutiny and torment.

In the Smithsonian Institute is a human skull 45,000 years old. Shape and alignment of the teeth indicate the owner of the skull was an upper-extremity amputee. India's *Rig-Veda*, an ancient epic poem over two thousand years old, recounts the tale of Vishpla, the warrior queen who lost a leg in battle, was fitted with an iron prosthesis, and returned to fight.

Ancient cultures were replete with amputee gods. The Peruvian jaguar god AiApec was an above-elbow amputee. Tezcatlitoca, the Aztec god of creation and vengeance, had lost his right foot. New Cah, a Celtic Irish god, was a left-arm amputee with a silver prosthesis replacing four fingers.

The prostheses of these cultures, described in both legend and art, used primitive materials available at the time. Egyptian mummies have been found with wrappings containing fibrous prostheses, probably created by the burial priests. In Greek mythology Pelops, grandson of Zeus, had an ivory shoulder fashioned by the goddess Demeter.

The earliest written record of an artificial limb can be traced to Herodotus who, in 500 B.C., wrote of a prisoner who escaped from the stocks by cutting off his foot and later replaced it with a wooden substitute.

Only sporadically did reports of limb replacements appear over the centuries. Pliny the Elder, in the first century, described a Roman general whose right arm had been amputated and replaced with an iron hand, fashioned to hold his shield and allow him to return to battle.

Little of note, scientifically or in any other field of human endeavor, was accomplished during the Dark Ages. The feudal system, with no central government or learning forum to help disseminate new findings and accomplishments, made development of new knowledge almost nonexistent in Europe. Few prosthetic alternatives were available at that time.

In the 1500s, before the advent of anesthesia, true cauterization, or the tourniquet, a surgeon had only thirty seconds to amputate a limb and three minutes to complete the operation—not a lot of time, and not enough to incorporate refinements. Many patients died of blood loss, or developed severe infections—even following finger amputation—after surgery. Not until the introduction of the tourniquet in 1674, and more extensive use of the ligature to tie off blood vessels and prevent fatal hemorrhage, did amputation become more of a curative procedure, rather than a last-ditch effort to save a life. Having a bit more time allowed surgeons the luxury of making amputations more functional, and prosthetists the opportunity to improve their prostheses.

Versatile French military surgeon Ambrose Pare championed the use of the ligature and abandoned the use of a favorite cauterization method, boiling oil, when he ran out during a battle. He invented, for a French army captain, a prosthetic hand called Le Petit Lorrain, a pride of watchmaker's skill with springs and catches, worn in battle. Despite the originality and effort poured into evolution of these prostheses, they were bulky, heavy, and only marginally functional.

Toward the beginning of the Renaissance, knights returning home could, if they could afford them, wear metal prostheses more cosmetic than functional to hide the disgraceful loss of a limb, a sign of weakness in battle. The iron arms designed by mercenary knight Gotz von Berlichingen in the early 1500s were the best examples of what could be achieved then. Mechanical masterpieces, the hands were suspended by leather straps. In a single-arm amputation, the finger joints could be moved into position with the good hand and relaxed with releases and springs.

Most of the prostheses made up to the 1800s were leg prostheses, designed and manufactured with great ingenuity. Still bulky and heavy until 1912, the age of aviation, when Englishmen Marcel and Charles Desoutter fabricated the first aluminum prosthesis, leg prostheses boasted hidden springs, concealed tendons, a smooth appearance, improved and better cushioned sockets, and clever mechanisms for creating foot movement and stability.

But despite these much heralded innovations, legs still had only two major functions: bearing weight and locomotion. This is a far cry from the myriad functional demands and complex needs of hands, for which so much usefulness hinges on feeling and dexterity, both of which have been beyond the reach of any artificial contrivance.

The American Civil War fueled the further development of prostheses. First of all, it vastly expanded the need: thirty thousand amputations in the Union Army alone. Second, the U.S. government made a major commitment, ongoing to this day, to provide prostheses to all needy veterans.

American prosthetists remained competitive and independent through the turn of the century. World War I did little to change this, since the war produced so few amputations in the United States (just over four thousand) compared with Great Britain (over forty thousand) and the rest of Europe (over one hundred thousand). As a result of these lopsided proportions, plus the financial depression in America, European prosthetists had reason to experiment and leap forward in their designs. When, in World War II, the number of American amputees caught up with European casualties, it was an impetus for the technology in this country to catch up. Great strides were made after 1946 when the U.S. Surgeon General brought a team of engineers and surgeons to Europe. Collaboration led to the formation of the American Orthotics and Prosthetics Association, with over 2,500 members—not enough to meet the need—through which could be developed ethical standards, scientific and educational programs, and improved relationships with health professionals.

In the last fifty years, research in both lower- and upper-limb prostheses has progressed further and faster than ever before in all of recorded history. The years 1983–1992 were declared the "decade of the disabled," partly in recognition of the emphasis leading up to the passage of the Americans with Disabilities Act, partly in recognition of technological advances designed to assist the disabled.

An orthosis is a brace or strengthening device; a prosthesis is an artificial limb. The approach to rehabilitation of a patient with an amputation involves consultations with an orthotist, prosthetist, physical therapist, counselor, and social worker—design, measurement, fabrication, fitting, and adjustment—all with input from the attending physician and the needs and desires of the patient.

The working end of a prosthesis is called a terminal device. Prosthetic replacements for lower extremities must meet the essential criteria of stability and sturdiness for weight bearing, as well as allow locomotion. Advances in materials, design, and fabrication have gone far in solving many of the major problems faced by lower-limb amputees.

Not so for those missing an arm or hand. The hand is simply too complex in its myriad functions for any artificial device to fill in. Upper-limb terminal devices are segregated into two basic categories: hooks for gross manipulation and hand surrogates for cosmesis.

Purely cosmetic hands and fingers, exemplified by the work of Jean Pillet, John Michael, and others, have a place both for traumatic as well as congenital losses. If only a fingertip is missing, a prosthesis can produce nearly normal appearance and function (gross grip, not feeling) of the affected hand. Finger absence may have a profound psychological effect. In some cultures, loss of a finger, or any hand deformity, may stimulate the individual to place the affected hand in a pocket or out of sight, producing a *de novo* hand amputation. The hand is there, but it might as well not be because it isn't used. The older prostheses were fabricated from polyvinyl chloride. This material stained easily and edges and joins were difficult to conceal, leading to poor patient acceptance.

Molded silicone, fabric reinforced, can be made using a lost wax technique. Durable, stain resistant, color matched to the normal hand, such prostheses can hide a wide range of abnormalities. The silicone may actually help to hydrate amputation stump tissues. Superb elasticity, intimate fit, thin margins, lifelike fingernails, and pleasing shape combined with artistic sensitivity help make such prostheses acceptable to a wide variety of amputees.

Hooks are no-nonsense appendages made of stainless steel. They can rotate, open, or close either using a harness attached to the opposite shoulder, or by electric motors. In the United States, around 70 percent of prosthesis users wear hooks. Outside the United States, especially in developing countries, there seems to be a preference for more cosmetically agreeable, hand-shaped prostheses, even though the additional cost is prohibitive.

In the early 1900s, arm prostheses consisted of a leather socket, which gave off a perspiration odor, a heavy steel frame, and an unsatisfactory terminal device; either a heavy leather or cotton-covered cosmetic hand; a clumsy mechanical hand with limited function; or a single hook for lifting and carrying. Dissatisfied, D. W. Torrance, a right-arm amputee, devised an improved, split hook with increased capabilities. But a hook is still a hook. Useful for gross or rough work, hooks have a proscribed functionality. For this, they do their job well. But they are not aesthetic, and are certainly not romantic.

In contrast are prostheses that are shaped and are intended to function like hands. The chief drawback of all synthetic hands is the complete absence of feeling. Despite this overwhelming limitation, a strange faith in the power of fantasy, or perhaps an infatuation with glitzy new technology has spurred the evolution of a steady stream of space-age designs. For those using body-powered prosthetic hands, the first preference for improvement is greater ability to hold both large and small objects. Users of electric hand prostheses have expressed a preference for bending of the fingers, and a thumb that can be positioned. Reduction in weight and energy expenditure continue to be top priorities.

Artificial hands powered by electrodes wired to muscles have been developed at Rutgers University; at UCLA; at Stanford's Rehabilitation Center; at Utah Arm Laboratories; at the Bioengineering Centre at Lothian Primary Care Trust in Edinburgh, Scotland; and elsewhere. Research at Strathclyde University, Glasgow, Scotland, has shown that children especially respond favorably to the fitting and use of an electric arm. Otto Bock Corporation in Germany has pioneered "smart prosthetics," which contain microprocessors allowing adjustment to individual use. Promotional literature has termed their "sensor hand" a "long hard stare into the future." Maybe. New technology will continue to excite, to dazzle, to encourage prodigious funding.

But, as orthotist/prosthetist Ernest Porter has cautioned: "In the future I think that prostheses will continue to develop, they will become faster and lighter and they will use less power. However, I don't think they'll ever compare to the actual human hand." The lack of sensibility in any manufactured, artificial prostheses hasn't stopped prosthetists from devising the most sophisticated—and expensive— mechanical gadgets imaginable.

Improved technology has been accompanied by increased expectations, and a form of cynicism has set in. Patients often exhibit an intense desire to have the prosthesis replace the lost extremity. Unrealistic movie scenarios, like the one showing Luke Skywalker getting an amputated hand replaced in *Star Wars*, haven't helped. Patients aren't happy or even satisfied with limitations; they want, and expect, perfection and for the most part grumble when they don't get it. With arm and hand amputations, perfection, or anything close to it, is simply too much to expect.

John Michael, certified prosthetist/orthotist and director of professional and technical services for Otto Bock, has written a refreshing disclaimer for electric arm prostheses, citing human qualities of success, independence, frustration tolerance, function—and wholeness—attainable by amputees with or without such a prosthesis. Independence lies in the attitude and efforts of the individual. As he suggested, people born without an arm segment have no loss to replace, only a different reality. Most people will, eventually, become comfortable with their bodies, without artifacts considered socially acceptable. A prosthesis cannot substitute for an already positive, strong self-image.

The broad sweep of history suggests that flirtation with technical mastery will continue without letup. The crush of patients desiring this technology will, in all likelihood, increase its availability and will only enhance the delivery of safe, quality service. Surgical techniques will improve as technology strengthens to meet demand. Most probably, the drugs will get better and safer, also in response to need.

Epilogue: Holography

Holographic principles might (also) govern interactions at the macrocosmic level of the entire universe. —Richard Gerber, M.D.

Take a hologram. Fragment it into fifty pieces. Take each piece and hook it to a special projection apparatus and project the contents of the piece. What do you think happens?

The entire hologram is reproduced, albeit with a little loss of detail. Each piece retains the information, the contents of the whole. It's the principle of holography.

Now, take a salamander. Teach it a simple behavior, like a reproducible response to a stimulus. Cut its brain into fifty pieces and transplant each piece into the brain of a salamander that hasn't learned the behavior (yes, it has been done). What do you think happens?

All fifty salamanders can now perform the behavior. They have "learned" the behavior by incorporating the transferred piece of brain into their own and assimilating its information, its contents.

Could the brain be like a hologram?

I'm certainly not the first to suggest that it is. Following Leith and Upatniek's 1965 *Scientific American* article on holography, scientists Bela Julesz and K. S. Pennington, Rutgers University physicists, explicitly theorized memory was stored in the brain as interference patterns similar to a hologram. Stanford neurophysiologist Karl Pribram was excited by the notion and, after experimental work on vision in monkeys, suggested other senses operated similarly. University of London quantum physicist David Bohm carried the concept even further. Indiana University anatomist Paul Pietsch wrote eloquently in support of the holographic theory of brain function. As have many others.

A number of observations reinforce this idea. Individuals who can hear with only one ear are able to pinpoint the location of a sound's origin. We can recognize a person we haven't seen in many years, even if time and aging have changed the person's appearance considerably. People whose visual cortex has been removed are still able to avoid stumbling over furniture in their path and "guess" at object identification, discriminating on visual tests with uncanny accuracy.

Trace, in the air, your signature with your elbow. You'll find you can do this easily, even though you've never done it before and you use a completely different set of muscles than the ones you use to write.

Karl Pribram suggested the amazing ability to transpose a skill to an entirely different scale or to a totally separate body part supported the concept of holographic principles at work.

If a brain contains six billion cells, each of which connects by axons and dendrites with others, the connectivity must be virtually incalculable. The primacy of individual axon-to-dendrite signals mediated by microscopic packets of neurotransmitters begins to break down when one considers the dazzling speed with which many complex, even automatic human actions are performed.

Pribram has theorized sensory perceptions are transformed into waves that crisscross the brain. Memory and sensory perception generate different kinds of waves that can interfere with one another, producing an interference pattern resembling a hologram. Much evidence suggests memory is stored throughout the body's tissues, not just in the brain.

Bodywork practitioners, like massage therapists and Rolfers, have noted tissue massage can sometimes evoke vivid memories and strong emotions, unbidden, often tied to traumatic life events. This repeated observation of "somatic recall" suggests information and memory are stored in the body's living cellular matrix.

Experiments by Karl Lashley showed a rat's ability to run a maze was not eliminated by removal of scattered or even large brain portions. The learned motor behavior seemed to be distributed throughout the brain. Pribram, one of Lashley's students, cited work by the Russian Nikolai Bernstein, who studied dance movements. In ingenious experiments, Bernstein painted white dots on dancers' black leotards and found, in motion pictures, that their movements resembled wave patterns comparable to those seen when a hologram is analyzed by Fourier calculus. To Pribram, this further supported the theory that the brain stores movements as wave patterns holographically and explains our rapid learning of complex physical tasks, such as virtuoso piano performance.

This vast oversimplification is, nonetheless, useful in explaining many unexplained observations. The hand is really a functional extension of the brain. In fact, the two are inseparable. When a concert pianist plays a piece without written music, with lightning speed, and with emotional content built into the performance, able to change any or all of the parameters in an instant, it is difficult to rationally believe otherwise.

Both the hand and the brain have evolved together, as a unit. Professor Michael Gazzaniga, director of the Center for Cognitive Neuroscience at Dartmouth College, theorized that brain hemisphere dominance was altered by humanity's extensive use of tools, which is a primary function of the right hemisphere, leaving the left hemisphere to focus on language. Our manipulative skills, along with our brains, have had eons to develop. The surpassing skill of the hand should come as no surprise.

The pioneering neurosurgeon and neurobiologist Wilder Penfield, after many years of doing brain surgery, asserted that no part of the cerebral cortex could be singled out as the focus of consciousness. In a similar way, no one part of the hand is the focus of function. Functional ability seems to be distributed throughout the hand. People find a way to do what they need to do, and the capacity of the hand to adapt to anatomic loss is both staggering and inspiring. In this ability, the hand is almost sentient.

In a way, it's easier to label the brain as holographic than it is the hand. We ascribe to the brain mysterious capacities, uncharted powers. It is, after all, the locus of control. Many believe "psi phenomena," like telepathy, teleportation, telekinesis, and levitation, all exist as potential capabilities in everyone, waiting for a phase shift in the brain's hologram to allow them to spring into common usage. Our ability to achieve this universally may be only a matter of time. Investigators like Stanislav Grof, originator of transpersonal psychology, believe we have these latent powers already, that every human being has access to all forms of consciousness.

And what of the hand? I believe it will evolve as the brain evolves, increasing its skills, its usefulness, in ways not now foreseeable. The time will surely come when full regeneration of critical areas of the brain and central nervous system are routine. Strokes will occur, but their consequences will be promptly reversible. And the hand will also benefit, since the urgent need for adaptation and rehabilitation will be eliminated.

Human beings, as a species, have apparently become comfortable with, or at least adjusted to, the mantle of protoplasmic reality imposed on us by evolution on planet Earth. True, we are using technology—in quantum physics, in the genome project, in medical imaging, in so many other areas—to push the envelope of our current limitations.

Yet despite this there is an underlying restlessness, a dissatisfaction with the frailties of the flesh. Our birthright is insufficient. Increasingly, we want to be part of a cosmic reality. If only we could understand it, fathom its scope, its boundaries, its implications. And, failing understanding, at least yearn intelligently.

Physicist David Bohm engaged in dialogues with the Indian spiritual master and mystic Krishnamurti. He postulated the existence of the Holomovement, the notion that physical reality, intricate and rich, is an undivided whole in a state of perpetual dynamic flux. He envisaged mindfulness on a grand scale, fulfilling the brain's potential as part of the natural order of things.

The hand and the brain will progress down the road of advancement "hand in hand," as humanity progresses. From a philosophical point of view, the holographic approach is consistent with deep, perceptive traditions of human experience. From a practical point of view, the scientific method, still the gold standard of modern problem solving, may well vindicate believers in as yet unmeasurable energies. We have our hands, and we have our minds. With them, we can, and will, do anything.

Recommended Readings

What has been written about each topic, from braille to sign language, from rheumatoid disease to carpal tunnel syndrome, would take up an enormous amount of space and, cumulatively, constitute its own library on the hand. Rather than cite the references I've consulted—thousands, over the years—I've suggested a few salient resources and Internet references as a way of acquiring more detail as well as additional bibliographies.

NAMES OF JOURNALS:

Am. J. Indust. Med.	=	*American Journal of Industrial Medicine*
Am. J. Phys. Med.	=	*American Journal of Physical Medicine*
Am. J. Psychol.	=	*American Journal of Psychology*
Ann. Int. Medicine	=	*Annals of Internal Medicine*
Ann. MBC	=	*Annals of the Burns and Fire Disasters*
Ann. Medit. Burns Club	=	*Annals of the Mediterranean Burns Club*
Ann. Rev. Med.	=	*Annual Review of Medicine*
Ann. Surg.	=	*Annals of Surgery*
Arch. Dermatol.	=	*Archives of Dermatology*
Arch. Neurology	=	*Archives of Neurology*
Behav. & Brain Sciences /BBS	=	*Behavioral and Brain Sciences*
Brit. Med. J.	=	*British Medical Journal*
Canad. J. Plastic Surg.	=	*Canadian Journal of Plastic Surgery*
JAMA	=	*Journal of the American Medical Association*
J. Burns	=	*Journal of Burns*
J. Inst. Gen. Med. Sci.	=	*Journal of the Institute of General Medical Science*
J. Invest. Derm.	=	*Journal of Investigative Dermatology*
Massage Therapy J.	=	*Massage Therapy Journal*

Orth. Clin. N.A.	=	*Orthopedic Clinics of North America*
Proc. Natl. Acad. Sci.	=	*Proceedings of the National Academy of Science*
Psych. Reviews	=	*Psychological Reviews*
Seminars in Occup. Med.	=	*Seminars in Occupational Medicine*
Southern Med. J.	=	*Southern Medical Journal*
Surg. Clin. N.A.	=	*Surgical Clinics of North America*

12: GESTURE

Fast, Julius. *Body Language.* New York: Pocket Books, 1970.

Frick-Horbury, D., and R. Guttentag. "The Effects of Restricting Hand Gesture Production," *Am. J. Psychol.* 111 (1998):43–62.

Morris, Desmond, et al. *Gestures: Their Origins & Distribution.* New York: Stein & Day, 1994.

Mulder, Axel. *Hand Gestures for HCI.* Burnaby, B.C., Canada: Simon Frazer Univ. School of Kinesiology, 1996.

Scheflen, Gilbert. *Body Language and Social Order.* Englewood, N.J.: Prentice-Hall, 1972.

Wolff, Charlotte. *The Psychology of Gesture.* London: Methuen & Co., 1945.

Zunin, Leonard, M.D. *Contact: The First Four Minutes.* New York: Ballantine Books, 1972.

13: SENSIBILITY AND TOUCH

Costello, Elaine. *Random House-Webster's American Sign Language Dictionary.* New York: Random House, 1997.

Courcey, Kevin, R.N. "Further Notes on Therapeutic Touch," http:www.quackwatch.com/01 QuackervRelatedTopics/tt2.html.

Finney, Dee. "Ancient Hand Signs," http://www.sweatdreams.com/hands.htm.

Gibson, J.J. "Observations on Active Touch," *Physiological Review* 69 (1962):477–490.

Hayward, Jonathan. "A Treatise on Touch," http://www.imsa.edu/ -ihayward/touch.html.

Heller, M., and W. Schiff, eds. *The Psychology of Touch.* Mahwah, N.J.: Lawrence Erlbaum Assoc., 1991.

Johansen-Berg, Heidi. "The Physiology and Psychology of Selective Attention to Touch," *Frontiers in Bioscience* 5 (11/1/2000):894–904.

Jones, Stanley. *The Right Touch.* Cresskill, N.J.: Hampton Press, 1988.

Krieger, Dolores. *Living the Therapeutic Touch.* New York: Dodd, Mead, 1987.

Lee, Richard. "Healing Hands Change Energy Fields," http://www.china
-healthways.com/TT.html.

Miles, Barbara. "Talking the Language of the Hands to the Hands,"
http://www.tr.wou.edu/dblink/hands2.htm.

Montagu, Ashley. *Touching: The Human Significance of the Skin*. New York:
Columbia Univ. Press, 1971.

―――. *Touching: The Human Significance of Skin*. New York: Harper &
Row, 1986.

O'Mathna, Dnal. *Therapeutic Touch: What Could Be the Harm? The Scientific Review of Alternative Medicine*. Philadelphia: Prometheus Books,
1998.

Research Council for Complementary Medicine. "Research on Touch,"
http://www.rccm.org.uk/masstouch.htm.

Rosa, Linda, R.N. "A Close Look at Therapeutic Touch," http://p6boss.
iroe.fi.cnr.it/misc/prano/htm.

14: LEFT-HANDEDNESS

Barsley, M. *The Other Hand*. New York: Hawthorn Books, 1967.

Blau, A. *The Master Hand*. Research Monograph no. 5. New York: American Orthopsychiatric Assn., 1946.

Corballis, Michael. *The Lopsided Ape*. New York: Oxford University Press,
1991.

Coren, Stanley. *The Left-Hander Syndrome*. New York: Free Press, 1992.

Cortese, J. "Lefthandedness," http://www.io.com/-cortese/left/southpaw.
html.

Geschwind, N., and A. M. Galaburda. "Cerebral Lateralization," *Arch.
Neurology* 42 (1987): 428–459; 42:564–578; 42:634–654.

Holder, M. K. "Famous Left-Handers," http://www.indiana.edu/-primate/
left.html.

Lam, Wai. "Hand Beliefs and Superstitions," http://pages.nyu.edu/-whl203/
leftv/language.htm.

Springer, S, and G. Deutsch. *Left Brain, Right Brain*. New York: W. H.
Freeman & Co., 1981.

Watkins, M. "Creation of the Sinister," http://hcs.harvard.edu/-husn/
BRAIN/vol2/left.html.

15: INTUITION AND PALMISTRY

Benham, William. *The Laws of Scientific Hand Reading*. New York: Knickerbocker Press, 1900.

Campbell, Edward. *The Encyclopedia of Palmistry*. New York: Berkley Publishing Group, 1995.

Cheiro. *Cheiro's Language of the Hand*. New York: Prentice Hall Press, 1987.

———. *Palmistry: The Language of the Hand*. New York: Gramercy Books (originally pub. 1894). New ed. New York: Random House, 1999.

Fitzherbert, Andrew. *The Palmist's Companion*. Metuchen, N.J.: Scarecrow Press, 1992.

Gettings, Fred. *The Hand: An Illustrated History of Palmistry*. Hamlin, N.Y.: 1965.

Gile, R., and L. Leonard. *The Complete Idiot's Guide to Palmistry*. New York: Aloha Books, 1999.

McCue, Donna. *Your Fate Is in Your Hands*. New York: Pocket Books, 2000.

Powers, Serena. *Aristotle's Treatise on Palmistry: 350 B.C.* London, 1738, quoted in Benham, 1900.

Reda, Sheri. "Your Life Is in Your Hands," *Conscious Choice* (March–April 1996).

Sorek, Batia. *The Complete Guide to Palmistry*. Hod Hasharon, Israel: Astrolog Publishing, 1998.

Sorell, Walter. *The Story of the Human Hand*. Indianapolis: Bobbs-Merrill, 1967.

Squire, Elizabeth D. *Fortune in Your Hand*. New York: Fleet Press, 1960.

———. *Palmistry Made Simple*. N. Hollywood, Calif.: Wilshire Book Co., 1979.

Wolff, Charlotte. *The Human Hand*. New York: Alfred A. Knopf, 1943.

———. *The Hand in Psychological Diagnosis*. London: Methuen & Co., 1951.

16: PHANTOM LIMB

Calford, M., and R. Tweedale. "Interhemispheric Transfer of Plasticity," *Science* 249 (August 1990):805.

Duncan, Lynette. "The Phantoms," http://webhome.globalserve.net/sds/thephantoms!.htm.

Flor, H., et al. "Phantom Limb Pain," 375 *Nature* (8 June 1995):482.

Frazier, S., and L. Kolb. "Psychiatric Aspects of Pain in the Phantom Limb," *Orth. Clin. N.A.* 1 (1970):481.

Jackson, J. Hughlings. "On the Comparative Study of Diseases of the Nervous System," *Brit. Med. J.* (August 17, 1889):355.

Jain, N., et al. "Neuronal Growth in the Brain," *Proc. Natl. Acad. Sci.*, 25 April 2000.

Leriche, Rene. *La Chirurgie de la douleur*. Paris: Masson, 1937.

McGrath, M. "Describing Phantom Limb Experience," http://serendip.
brynmawr.edu/bb/neuro/neuro98/202s98.

Melzack, Ronald. "Phantom Limbs," *Scientific American* 266 (April
1992):220.

———. "Pain: Past, Present and Future," http://www.alternatives.com/
raven/cpain/melzack2.html.

Mitchell, Silas W. "Phantom Limbs," *Lippincott's Magazine of Popular Literature* 18 (1871):561–569.

———. *Injuries of Nerves and Their Consequences.* Philadelphia: J. B. Lippincott, 1872.

———. *The Autobiography of a Quack and the Case of George Dedlow.* New
York: The Century Company, 1900.

Ramachandran, V. S. *Phantoms in the Brain.* London: Fourth Estate, 1998.

Saadah, E. S. M., and R. Melzack, "Phantom Limb Experiences," *Cortex* 30
(1994):479.

Sherman, R. "Established Treatments of Phantom Limb Pain," *Am. J. Phys.
Med.* 59 (1980):232.

Sunderland, Sydney. *Nerves and Nerve Injuries.* New York: Churchill Liv-
ingstone, 1978.

17: SKIN GRAFTING

Artz, C. P. "Historical Aspects of Burn Management," *Surg. Clin. N.A.* 50
(1970):1193.

Burke, J. F., I. V. Yannas, and W. C. Quinby. "Successful Use of Physiologically Acceptable Artificial Skin," *Ann. Rev. Med.* 38 (1987):413–428.

Codina, T., P. del Caz, and M. Safont. "A Synthetic Skin Substitute in the
Treatment of Burns," *Ann. Medit. Burns Club* 8 (September
1995):1–5.

Davis, A. "Supported Basic Research on Skin Replacement," *J. Inst. Gen.
Med. Sci.* (1999).

Demling, R., L. DeSanti, and D. Orgill, "Skin Substitutes in Burn Management: Historical Perspective," http://www.burnsurgery.org/
Modules/skinsubstitutes/sec3.htm.

Hansen, S., et al. "Using Skin Replacement Products to Treat Burns and
Wounds," http//www.woundcarenet.com/advances/clinicalmgmt/
wct224.htm.

Heimbach, D., et al. "Artificial Dermis for Major Burns," *Ann. Surg.* 208
(1988):313–320.

Phillips, Tania. "New Skin for Old," *Arch. Dermatol.* 34 (1998):344–349.

Polywatis, G. E. "Historical Landmarks in the Treatment of Burns," *Ann. MBC* 2 (1989):1–5.

Prasanna, M., and S. Kuldeep. "Early Burn Wound Excision and Skin Grafting," *J. Burns* 2 (1984):1–5.

Pulaski, E. "Thermal Burns," http://www.armymedicine.army.mil/history/booksdocs/KOREA/recad1/ch6-4.htm.

Revis, D., and M. Seagel. "Skin and Grafts," http://www.emedicine.com/plastic/topic392.htm.

Sher, B., et al. "The Reconstitution of Living Skin," *J. Invest. Derm.* 81 (1983):2s–10s.

Sneve, H. "Treatment of Burns and Skin Grafting," *JAMA* 45 (1905):1–8.

Thoma, A. "Plastic Surgery and Greek Mythology," http://www.pulsus.com/Plastics/03 02/thom ed.htm.

Weiss, H., E. Wertheym, and R. Shafir. "Synthetic Skin Substitute for Pediatric Burns," *Ann. Medit. Burns Club* 6 (June 1993):1–4.

18: CARPAL TUNNEL SYNDROME

Armstrong, Thomas. *Biomechanical Aspects of Upper Extremity Performance & Disorders*. Dept. of Environmental Health monograph. Ann Arbor: Univ. of Michigan, August 1984.

Feldman, R., R. Goldman, and M. Keyserling. "Peripheral Nerve Entrapment Syndromes," *Am. J. Indust. Med.* 4 (1983):661–681.

Flynn, J. Edward. *Hand Surgery*. Baltimore: Williams & Wilkins, 1966.

Silverstein, B., L. Gine, and T. Armstrong. "Carpal Tunnel Syndrome," *Seminars in Occup. Med.* (September 1986):213–221.

Spinner, Morton. *Injuries to the Major Branches of the Peripheral Nerves of the Forearm*. Philadelphia: W. B. Saunders Co., 1978.

19: RHEUMATOID DISEASE

(R.A. = Rheumatoid Arthritis)

ACR. "Guidelines for Management of R.A.," *Arthritis and Rheumatism* 39 (May 1996):713–722.

Annals of Internal Medicine. "R.A.:Treat Now, Not Later," editorial, http://www.acponline.org/journals/annals/15apr96/arthedit.htm.

Brennan, Barbara. *Hands of Light*. New York: Bantam Books, 1987.

Dossey, Larry. *Recovering the Soul*. New York: Bantam Books, 1989.

Easterbrook, Mark. "Can You Pray Your Way to Health?" http://www.pastornet.net.au/imm/spir0019.htm.

Flatt, Adrian, M.D. *The Care of the Rheumatoid Hand*. St. Louis: C. V. Mosby, 1963.

Green, Saul. "Facts or Fraud," speech delivered to Harvard Alumni Association, New York, May 9, 2000.

Hansen, Lars. "Fish Oils and Rheumatoid Arthritis," International Health News Database, http://www.oilfpisces.com/rheumatoidarthritis.html.

Horstman, J. "More Than Medicine," *Arthritis Today*, 12 May 2001.

ICBS, Inc., http://www.holistic-online/Remedies/Arthritis/arth_RA_prayer.htm.

Kelleher, J. "Prayer: A Healing Gift," http://health.infospace.com.

Last, Walter, "Arthritis and Rheumatism," http://greenmini.net.au/-wlast/arthritis.htm.

Matsen, S., M.D. "Rheumatoid Arthritis," http://www.orthop.washington.edu/BoneJoint/rzzzzzzzl_2html.

Matthews, D., S. Marlowe, and F. MacNutt, "Effects of Intercessory Prayer on Patients with R.A.," *Southern Med J.* 93 (December 2000).

Posner, Gary, M.D. "An Examination of the Media Coverage of a Prayer Study-in-Progress," http://members.aol.com/garypos/prayerstudyinprog.html.

Ross, Clare, R.N. "A Comparison of Osteoarthritis and R.A.: Diagnosis and Treatment," *The Nurse Practitioner*, September 1997.

Univ. of Washington, Orthopedics and Sports Medicine, "Rheumatoid Arthritis," http://www.orthop.washington.edu/arthritis/rheumarth.

20: THUMB APLASIA: MISSING THUMBS

Damasio, Antonio. *Descartes Error*. New York: Grosset/Putnam, 1994.

Gupta, A., S. Kay, and L. Scheker, eds. *The Growing Hand*. London: Mosby, 2000.

Kramer, Anthony. *Hands Up*. New York: Macmillan, 1983.

Morris, P., and S. Morris. *The Panda's Thumb*. Chicago: Athro, Ltd., 2000.

Peacock, Erle E., in J. E. Flynn, ed., *Hand Surgery*. Baltimore: Williams & Wilkins, 1966.

Penfield, Wilder. *The Cerebral Cortex of Man*. New York: Hafner Co., 1968.

Pribram, Karl. *Languages of the Brain*. Englewood, N.J.: Prentice-Hall, 1971.

Sylwester, Robert. "A Celebration of Neurons." N.p.: Association for Supervision & Curriculum Development, 1995.

Wilkins, W., and J. Wakefield. "Brain Evolution and Neurolinguistic Preconditions," *Behavioral and Brain Sciences* 18 (1995): 161–226.

American Academy of Orthopedic Surgeons. *Atlas of Limb Prosthetics*. St. Louis: Mosby, 1981.

Broomfield, Mark. *A Guide to Artificial Arms*. Glasgow, Scotland: University of Strathclyde, Reach, 2000.

CNN Interactive. "Transplant Patient Enjoys New Hand," (AP), Louisville, Ky., 26 January 1999.

Doshi, R., et al. "The Design and Development of a Gloveless Endoskeletal Prosthetic Hand," *J. of Rehabilitation Research and Development* 35 (October 1998):388–395.

Entin, M., M.D. "The Story of Replantation and Microsurgery," *Canad. J. Plastic Surgery* (1999):249–254.

Harderosian, A. D. "Medicinal Leeching, Past and Present," *Thrombosite: Newsletter of Pharmaceutical Information Associates, Ltd.* 1(3) (1999):1–12.

Klopsteg, P., and P. Wilson. *Human Limbs and Their Substitutes*. New York: Hafner Co., 1968.

Kurzman, Steven. "Anthropology and Prosthetics: History," *Physical Medicine and Rehabilitation of North America* 2 (1991).

Langdorf, M., and Z. Kazzi. "Replantation," http://www.emedicine.com/emerg/topics502.htm.

Levy, S. W., M.D. "A Prosthetic Primer," *Biomechanics Magazine* (April 1999):45–54.

Mooney, R. L. *Surviving an Amputation*. Los Angeles Mutual Amputee Aid Foundation, 2001.

Pillet Hand Prosthesis, http://www.orthopedie.com/php_2/pillet_faq.htm.

Prosthetics history, http://www.ampulove.com/ampinformation/proshistory/proshistory.htm; http://www.nupoc.northwestern.edu/prosHistory.shtml.

Rank, B., A. Wakefield, and J. Hueston. *Surgery of Repair as Applied to Hand Injuries*, 4th ed. New York: Churchill-Livingstone, 1973.

Resnick, Jeffrey. "Historical Perspectives on the Art and Science of Prosthetics," http://www.service.emory.edu/-jreznic/TechShatten.htm.

Stix, Gary. "Phantom Touch," *Scientific American*, October 1998.

Transplantation, http://www.library.nuigalway.ie/webpath/tutorial/transpl/transpl.htm.

Vizard, Frank. "Bodies by Design," *Popular Science*, October 1999.

Daily, C. "Somatic Recall: Soft Tissue Holography," *Massage Therapy* J. 34 (1995).

Eccles, John. *Evolution of the Brain*. London and New York: Routledge, 1989.

Gerber, Richard, M.D. *Vibrational Medicine*. Santa Fe, N.M.: Bear & Co., 1996.

Grof, Stanislav. *The Holotropic Mind*. New York: HarperCollins, 1992.

Harth, Eric. *Windows on the Mind*. New York: William Morrow, 1982.

Heckman, Philip. *The Magic of Holography*. New York: Atheneum, 1986.

Kasper, J., and S. Feller. *The Hologram Book*. Englewood, N.J.: Prentice-Hall, 1985.

Keepin, Bill. "Lifework of David Bohm: River of Truth," http://www.vision.net.au/-apaterson/science/david bohm.htm.

Pietsch, Paul. *Shufflebrain*. New York: Houghton-Mifflin, 1981.

Pribram, Karl. *Languages of the Brain*. Englewood, N.J.: Prentice-Hall, 1971.

―――. "Folism Could Close the Cognition Era," *APA Monitor* 16 (1985): 5–6.

―――. *Brain and Perception*. Mahwah, N.J.: Lawrence Erlbaum Assoc., 1991.

Prideaux, J. "Comparison Between K. Pribram's Holographic Theory and More Conventional Models of Neuronal Computation," http://www.acsa2000.net/bcngroup/iponkp/.

Prigogine, I. *Order Out of Chaos*. New York: Bantam Books, 1984.

Reich, Wilhelm. *Cosmic Superimposition*. New York: Farrar, Straus & Giroux, 1973.

Talbot, Michael. *The Holographic Universe*. New York: HarperCollins, 1991.

Wilber, K. *The Holographic Paradigm and Other Paradoxes*. Boston: Shambhala, 1985.

ADDITIONAL SELECTED REFERENCES

Broker, Stephen. "Hominid Evolution," http://www.yaie.edu/ynnu/curriculum/units/19/9/6//9.06.02.X.html.

Bower, Bruce. "I.Q.'s Evolutionary Breakdown," *Science News* 147 (April 1995):220–222.

Donald, Merlin. "Precis of Origins of the Modern Mind," *Behav. and Brain Sciences* 16 (1997):737–791.

Gibson, K., and T. Ingold. *Tools, Language and Cognition in Human Evolution*. U.K.: Cambridge Univ. Press, 1993, pp. 230–250.

Johanson, Leroza Carey. "Lucy Turns 25," *ASU Research*, Fall 2000.

Jourdain, Robert. *Music, the Brain and Ecstasy*. New York: William Morrow, 1998.

Kaas, Jon. "What, If Anything, Is S.I.?" *Psych. Reviews* 63 (January 1983): 206–230.

MacNeilage, P. "The Frame/Content Theory of the Evolution of Speech Production," http://www.cogsci.soton.ac.uk/pps/Archive/pps.macneilage.html.

Napier, John. "The Evolution of the Hand," *Scientific American* 207 (December 1962):56–62.

———. *Hands*. Princeton, N.J.: Princeton Univ. Press, 1963.

Paracelsus. *Four Treatises*. Baltimore: Johns Hopkins Univ. Press, 1941 (originally pub. 1538).

Posner, H. "BBS Precis," http://www.arts.uwaterloo.ca/~acneyne/donald/html.

Schilder, Paul. *The Image and Appearance of the Human Body*. New York: International Univ. Press, 1974.

Spinner, Mort. *Kaplan's Functional and Surgical Anatomy of the Hand*. Philadelphia: J. B. Lippincott Co., 1978.

"The Genesis of Mind," http://www.marxist.com/science/genesisofmind.html.

Wilson, Frank. *The Hand*. New York: Vintage Books, 1998.

Acknowledgments

The heroes and heroines of this book are my patients. The success of these ordinary folks faced with immense challenges is a tribute to their humanity. I salute them.

Writing this book is a new venture for me, which I could not have successfully completed without lots of help. Luckily, talented people surrounded me who offered professional guidance when I most needed it.

My editor, Erika Goldman, has been a joy to work with. She demonstrated, from the beginning, a fine sensitivity to the complexities inherent in organizing the vast subject with which I wrestled. Her incisive editorial comments challenged my insecure clinging to my "golden prose." The book is better for her educated whittling. Her assistant, John Parsley, has been more than kind in coercing me to adhere to deadlines. They make a fine team. To both of them, I offer my heartfelt thanks.

Thanks, too, to my agent, Andrew Stuart. Andrew found me at a writing conference, spirited me away, and then spent the next year shaping this volume for publication. Exhibiting the patience of Job and the humor of Rodney Dangerfield, he guided me in the mysterious workings of the publishing industry. I might have found a way to get the book into print without his help, but it certainly would have been more of a struggle, and a lot less rewarding.

Along the way, my family, friends, and colleagues have had an opportunity to critique individual chapters. Their comments (usually kinder and gentler than those offered by my writing classmates) were to the point and invaluable to me. I could not possibly compile a complete list—there were too many over the four years of gestation of this book and I didn't keep track—but I am grateful for the time spent and the well-intended comments. A few names do stand out and deserve special recognition: Jennifer Schneider, Peter Kay, Kimba Arem, Dan Dudley, Bert and Betty Feingold, Dan Levinson, Bob Powers, Scott Hadley, Burt Schneider, Jacquie Brailey, Adrienne O'Hare, Sylvia Arem, Keith and Val Arem, Joel Arem.

One person in particular deserves my deepest expression of awe, respect, and appreciation. Meg Files, a fine writer, the author of several books, and a superb writing teacher, was unbelievably supportive of me as my originally vague intentions slowly coalesced. I literally could not and would not have completed this project without her inspiration and help. Thank you, Meg, for being there, always.

At the top of the list, where she always resides, is my wife of thirty-seven years, Cindy. My soul mate, my best friend, she is and always has been the wind beneath my wings. We share more than could possibly be expressed here. Thank you, Cindy, for your love, your graceful strength, your sustenance, your tolerance.

Index

germ theory, 227
Geschwind, N., 180
gestures, 2, 155–60
    to augment communication, 155, 156, 157
    culture-dependent, 2, 157–58
    in dance, 159, 170
    derivation of the word, 156
    ethnic and cultural differences in use of, 2, 157–58
    hands' role in, 159
    heiratic, 159
    iconic, 156
    by infants, 155
    as international idiom, 157
    origins and meanings of common, 157
    as preverbal language, 155, 172
    sign language, 2–3, 156, 159–60
    symbolic meaning of, 2
Gibson, J. J., 162
glomus apparatus, 74
Goldberg, E., 179
gold therapy, 108, 213
Golgi apparatus, 162
Greece, ancient, 185–86, 236
Green, Dr. Elmer, 218, 220
Green, Saul, 217
Grobstein, Paul, 199
Grof, Stanislav, 246
guided imagery, 216
Gypsies, 185

Hagen, Johann (Indagine), 190
Hallam, Clint, 233–34
Halstead, William, 227
hand-clapping, 157
handedness, 117
    historical view of, 174–76
    left-handedness, see left-handedness
    two hemispheres of the brain and, 177–80
hand line interpretation, see palmistry
hand-mouth motions:
    in infants, 223–24
    in utero, 222
handprints of ancient civilizations, 183
handshake, 157
Hands of Children, The (Speer), 188

hand surgery:
    amputation of fingers, see amputations, of fingers
    on atypical tuberculosis patient, 147–49
    authorization for, 120, 122–25, 130–31
    for carpal tunnel syndrome, 97–98
    "cutting surgeons," 39
    difficult judgments made during, 15, 77–78, 81–84, 113
    first intervention, 4
    metacarpal transfer, 46–47, 49–50, 55–56
    operating room, see operating room
    on rattlesnake bite patient, 27–30
    replantation, see replantation
    for rheumatoid arthritis, 109, 112–14
    risks of, 50, 65
    safe time limit for, 4, 15, 78, 127
    scrubbing for, 77
    skin grafts, 62–72, 78, 79–80, 83, 84–85, 203–204
    on streptococcus patient, 14–16
    thumbs, to make a child, 117–32
    uncontrolled bleeding and, 126
haptic touch, 162
Harlow, Harry, 164
Haycraft, John, 231
Healing Words (Dossey), 218
heart transplantation, 228, 232
Hebb, Donald, 199
Heberden's nodes, 36, 43
heiratic gestures, 159
hemlock, 227
hemp, 227
henbane, 227
Henson, Jim, 9, 10
Herodotus, 236
Hinduism, 170
Hippocrates, 177
hirudin, 231
histiocytes, 146
history of the hand, informal, 155–247
HLA antigens, 232
Hogarth, William, 158
holistic approaches to rheumatoid disease, 214–21
holography, 198, 243–47
Holomovement, 246

Homer, 185
homo erectus, 175
homo habilis, 175
homosexuality and left-handedness, 181
homunculus, 196, 223
hooks as upper-limb terminal devices,
    239–40
Horus, Egyptian god, 176
hospitalization, 137, 138, 140
    for skin grafts, 65
human energy field (HEF), 167–71, 218
humors, four, 230
Hunt, Dr. Valerie, 168
Hunter, Thomas, 227
hydroxychloroquine, 213
hyperpyrexia, 70
hypnosis, 43
hysterical conversion reaction, 41–44

immunologic rejection:
    anti-rejection drugs, 204, 228, 232–33,
        234
    of cadaver skin grafts, 204
    of transplants, 232–34
Imuran, 213
Incas, 169
index finger, 12, 15
    functions of the, 25–26
    long finger made to work like an, 26,
        29
    loss of, from rattlesnake bite, 22–26
    making a thumb from an, 121–22,
        127–28
India, 168–69, 170, 176, 230
    origins of palmistry, 184
    Rig-Veda, 236
infants:
    gesture by, 155
    hand-mouth coordination
        movements, 223–24
    touch and:
        emotional development, 163
        premature infants, 164
inosculation, 67
insurance coverage, 140
    carpal tunnel syndrome and, 92–93,
        96, 98–101
Integra, 204
integrative approach to rheumatoid
    disease, 214–21

intercessory prayer, 218–20
intuition, 187–88
Inuits, 163, 175
involuntary nervous system, 53
ISSSEEM (International Society for the
    Study of Subtle Energies and
    Energy Medicine), 217–18
Iwerks, Ub, 45

jactitation, 48, 51
Japan and Japanese, 169
Joan of Arc, 177
Johannesson, Alexander, 156
Joshi caste of northwest India, 184
*Journal of the American Medical
    Association*, 171
Julesz, Bela, 243
Jung, C. G., 188
juvenile rheumatoid arthritis, 211

Kabbalah, 170
Kathy (patient pseudonym), 133–42, 228
Kelley, L. H., 175
kerosene plasters, 216
kidney transplantation, 227, 228, 232
Kilner, Dr. William, 171
knights, metal prostheses worn by, 237
Koobi Fora, Kenya, 174–75
Krieger, Dolores, 167, 218
Krishna, Lord, 184
Krishnamurti, 246
Kunz, Dora, 167

Labastide caves, 183
Lane, Harlan, 171
Lascaux grotto, caves of, 183
Lashley, Karl, 244–45
Last, Walter, 216
Lavoisier, Antoine, 168
"laying on of hands," 167, 168–71, 186
learning disabilities, 180
Lee, Richard, 171
leeching, medicinal, 230–32
left-handedness, 173–82
    chemical environment of the fetus
        and, 180–81
    denigration of, 173–74, 176–77,
        181–82
    devices favoring right-handed use,
        181

New Cah (Celtic Irish god), 236
New Testament, 186
noradrenaline (norepinephrine), 53, 166
NSAIDs (nonsteroidal anti-
    inflammatory drugs), 213, 214
Nylen, 229

occultism, 189
omega-3 oils, 216
omega-6 oils, 216
"On the Lines of the Hands," 185
operating room:
    banter in the, 14
    comraderie in, 27
    reverse precautions in, 147
opium, 227
Optimal Wellness Center, 214
organ transplantation, 227–28, 232, 233
oriented touch, 162
oscilloscope, 38
osteoarthritis, 35
Otto Bock Corporation, 240–41

Paget, Sir Richard, 156
pain medication, post-operative, 28, 69
palisading, 146
palmistry, 170, 183–91
    in ancient civilizations, 183, 184–87
    in the Middle Ages, 187
    Wolff and, 187–88, 189, 190
Paracelsus, 170, 186
paralysis of muscle caused by hysterical
    conversion reaction, 41–44
"parasympathetic" nervous system, 53
Pare, Ambrose, 192, 227, 237
Pelops (Greek mythological figure), 236
Penfield, Wilder, 3, 196, 197, 223, 245
Pennington, K. S., 243
pericarditis, 210
PET scans, 224
phantom limb, 47–56, 192–203, 223–25
    aplastic phantoms, 222–23, 225
    the brain and, 193, 196–201
    cramping and spasm, 195–96
    faulty amputation and, 194–95, 201
    historically, 48
    in history and literature, 192–93
    interruption of sensory pathways
        between limb and cerebral cortex
        and, 199–200

origination in the brain of, 193
persistence of, 49
    faulty amputation and, 194–95
prostheses and, 48
sensations versus pain, 195, 201
treatment of pain, 202
triggers that can intensify, 48, 195
younger children and, 196
physical therapy, 14, 16, 33, 137
Physiognomy and Palmistry (Pythagoras),
    185
Picasso, Pablo, 226
Pietsch, Paul, 244
pigskin for skin grafts, 203
Pillet, Jean, 140, 141, 235, 239
pinch power, 26, 109
placebo effect, 193
Plaquinyl, 213
Plastic Surgery Research Council, 150
Plato, 186
pleuritis, 210
Pliny the Elder, 186, 236
Pons, Tim, 199
Porter, Ernest, 241
positive thinking, 202
poultices for rheumatoid disease,
    216–17
prana, 170
prayer, intercessory, 218–20
Pribram, Karl, 243, 244, 245
prostheses, 51, 140, 141, 234–41
    in ancient civilizations, 236
    expectations of, 241
    first aluminum leg, 237
    history of, 236–38
    leg, 237–38
    phantom pain and, 48
    "smart," 240–41
    upper-limb, 239–41
        hand surrogates for cosmetic
            purposes, 239, 240–41
        hooks, 239, 240
psi phenomenon, 246
psychological causes of hand problems,
    51–52
    hysterical conversion reactions, 41–44
psychotherapy, 43–44, 51
Puri tribe of Brazil, 155
Pythagoras, 170, 185
Pythagorean Table of Opposites, 176

sensory capacity of the hand, 3, 92
sensory exam, 60–61
septic emboli, 10, 13
Set, Egyptian god, 176
sexual development, sense of touch and, 164
Shealy, Norman, 218
sica, 210
Sicilians, use of gesture by, 158
Sigismund, Holy Roman Emperor, 185
sign language, 2–3, 156, 159–60, 224
silicone implants to keeps bone end apart, 113
Sjogren's syndrome, 210
skin disorders, 164
skin grafts, 62–72, 78, 79–80, 83, 84–85, 203–4
   with animal skins, 203
   from cadavers, 203–4
   color change (inosculation), 67
   cosmetic appearance of, 88
   future of, 204
   history of, 78, 203
   matching growth of child, 88
   scar contraction, 64, 66
   with synthetic products, 204
   "taking" of, 84–85
   thickness of, 80
   window of opportunity for, 64
skin sensitivity to touch, see touch, sense of
skin substitute, search for a, 63
small finger, 12, 15
   carpal tunnel syndrome symptoms and, 91
Smithsonian Institution, 236
smoking and rheumatoid disease, 215
snake bites, poisonous, 19–34
somatic recall, 244
Sorell, Walter, 186, 190–91
Soria, Anna (patient pseudonym), 73–88
Southern Medical Journal, 218
Speer, Julius, 188
Sperry, Robert, 178–79
Squire, Elizabeth, 185, 189–90
Stanford Rehabilitation Center, 240
Stepanov Dance Notation, 159
stool guaiac test, 65
Stowburn, Marva (patient pseudonym), 35–44

Strathclyde University, 240
streptococcus, 9–18
   diagnosis of, 10
   hand surgery of patient with, 14–16
stroke, 246
Sulfamylon, 60
sulfasalizine, 213
Sushruta (Indian surgeon), 78
Swanson, Al, 113
"sympathetic" nervous system, 53–54
synapse, 198
synovitis, 210
synovium, 94, 95, 146, 148, 206, 208–9, 210
   as target of rheumatoid disease, 211–12

Talmud, 177
Tamai, Dr. S., 229
Taoism, 169
Tao Te Ching, 168
Taurog, Dr. Joel, 219, 220
Teichenbach, Count Wilhelm von, 170
telekinesis, 246
telepathy, 246
teleportation, 246
temperature regulation by the hands, 78–79
tendonitis, 146, 208–9
tendons, 93–94
   extensor, 25, 26, 47, 110, 111, 228
   in finger amputations, 136
   flexor, 47, 206, 207, 228
   function of, 5
   overuse of, 94, 95
   scarring, effects of, 5
   synovium surrounding the, 94, 95, 146, 148
Tezcatlitoca (Aztec god), 236
thalidomide, 118
Theosophical Society in America, 167
Therapeutic Touch (TT), 167–68, 170–71
three-fingered hand:
   of cartoon characters, 45, 56
   metacarpal transfer, 46–47, 49–50, 55–56

About the Author

ARNOLD AREM, M.D., Clinical Associate Professor at the University of Arizona College of Medicine and Clinical Associate in Surgery at the University of New Mexico, has been a hand surgeon for more than twenty years. A former vice president of the Tucson Holistic Health Association, Dr. Arem is an international lecturer and educator. A published medical illustrator as well, he has served as an industrial consultant for companies such as IBM and Marion-Dow Laboratories. In 1984 he developed "hand-outs," unique hand illustrations printed in pads and used for patient education. See *http://www.handouts.cc*. He is the author of numerous scientific articles. This is his first book.